In Nature's playground,
characters jog around,
hiding and peeking,
in all innocent fun,
but every so often,
a fight breaks out...

GENOMIC QUIRKS

The Search for Spelling Errors

Dedicated to the patients on whom
the stories in this book are based

ISBN 978-93-5265-745-2 (paperback)
ISBN 978-93-5265-743-8 (hardback)
ISBN 978-93-5265-744-5 (ebook)
ISBN 978-81-9257-074-7 (India paperback)

Published by IIScPress & Strand Life Sciences
www.iisc.ac.in/outreach/publications/iisc-press/
www.strandls.com
First Edition, Jan 2018

Contents

Preface

Imagine a physician trying to diagnose a patient's condition. The patient's symptoms appear mysterious, leaving the doctor stumped and the patient helpless. How can the doctor dig deeper to obtain a diagnosis?

Doctors use various tools to dig deeper. They use X-rays, MRI scans, CT scans and ultrasound scans to view our internal organs. They even view individual cells from these organs under a microscope by drawing a few drops of body fluid or by conducting surgical biopsies. But none of these helps when the cause of disease lurks deeper—at a molecular level.

In and around each of our tens of trillions of cells are a large number of molecules. Reactions between these molecules keep us going. Not surprisingly, some of our health problems are consequences of these molecules not behaving as we expect them to. However, probing these molecules for abnormal behavior is not easy for two simple reasons: they are really tiny, and they are way too many in number.

In this soup of molecules, the pride of place is occupied by the *Genome*. Let us spend a few minutes on the genome before we return to the plight of our doctor and his/her patient.

The genome is our parents' gift to us. Many other molecules in our body are derived from the genome in an indirect way, making it the master molecule. As a molecule (or actually a collection of molecules), it has its own distinctive form and function. Nevertheless, in our simple-minded view, it is just a book of characters—six billion characters that allow us to resemble our parents among a vast array of other functions.

The specific characters in your genomic book may be different from those in mine. And therein lies a fundamental question. What do these six billion characters tell us about ourselves? Can they predict the future trajectory of our health? Can they predict if we will grow up to be good athletes, scientists or orators?

The answer is less dramatic than we might hope. For instance, identical twins are born with supposedly identical genomes. Yet, one twin may develop autism, schizophrenia, diabetes or cancer while the other does not. This serves as a reminder that genomic characters are not the sole determiners of our fate. Rather, they act in concert with various environmental factors and with sheer random chance. Even when they act by themselves, they tend to do so collectively in large groups, making it hard for us to predict their combined effect. For instance, the height of an individual is probably determined by hundreds if not thousands of genomic characters working in concert.

Yet, there are a few quirky characters which drive home their agenda almost single-handedly. *A few* actually means a few hundred thousand, but nevertheless a small fraction of the entire pool of six billion characters. Typographical errors in such characters, though rare, can have a dramatic effect. Indeed, the root cause of our patient's illness may well lie in such an error. How can our doctor ever pinpoint this error in the midst of six billion characters?

For that, we have some remarkable technological advancements in recent years to thank. These make it possible for us to read the tiny genomic book of our patient, at modest costs. Indeed, even five years ago, this wouldn't have been possible, and this book couldn't have been written.

Returning to the plight of our doctor and her patient: today, the doctor can simply order a *genome sequencing* test if the suspected cause of illness is an abnormal genomic character. A request for this test, accompanied by some saliva or blood or biopsy tissue from the patient, arrives at a specialized genomics laboratory, such as the one at Strand Life Sciences. A few days later, the doctor receives a report from the laboratory describing any abnormal genomic characters found. Simple, isn't it?

Well, what happens in the interim is anything but simple. A very elaborate quest for the offending character is conducted in the laboratory. Since requests from hundreds of patients might arrive at any given point, several such quests are run simultaneously. The complexity and gravity of these quests then makes the laboratory seem very much like a war room.

In this genomics war room, molecular biologists in lab coats, or their robot equivalents, scurry around transferring samples from one receptacle to another. In this process, genomic molecules are extracted from the patient's sample, chopped into small fragments, and subjected to a multitude of manipulations. Eventually, the genome sequence comes out, but rather indirectly—in the form of millions to billions of tiny fragment sequences that amount to gigabytes of data. High-performance computers then piece these fragments back together, like a jigsaw puzzle. A comparison between this assembled jigsaw and that for a normal person yields a long list of *variants*, i.e., characters that are likely suspects. Computers then scour vast amounts of biomedical literature for information that helps zoom into the most relevant of these suspects, typically a handful in number. Finally, trained geneticists rack their brains to identify the ones that hold a clue to the patient's condition. If all goes well, one or two clear candidates emerge and are reported back to the doctor. But things don't always go well.

Many parts of the genomic jigsaw puzzle look very similar, forcing bioinformaticians to strain and spot subtle differences in order to make progress. Variants about which very little is known are found sometimes, forcing geneticists to go out on a limb and make educated guesses to establish culpability. No variants of note are found sometimes, leaving everyone scratching their heads and prompting a relook at the gigabytes of data at hand. Some variants are hidden more subtly under these mounds of data and need new algorithms to uncover. Inconsistencies on account of noisy data arise occasionally and have to be reconciled. Anxious doctors and patients call in the meantime enquiring for results. These results sometimes call for difficult decisions to be made. All in all, a recipe for frequent huddles, frustrating roadblocks, and ecstatic aha moments.

The nine stories in this book provide glimpses into this genomics war room. Each story deals with a distinct patient suffering from a distinct illness, the root cause of which lies in a distinct genomic quirk and often requires a distinct algorithm to unravel.

The first story is my personal one and deals with a relatively minor quirk of vision—color blindness. Serious illness is never a pleasant topic to discuss, hence the decision to use this relatively minor condition to provide a friendly introduction to the genome before we encounter graver conditions in subsequent stories. Tracking down the cause of my color blindness takes us through the science of color perception and the vagaries of some cut-and-paste events that happen to the genome as it passes down the generations from parents to children.

The second story deals with a family in which several members face loss of central vision in their 30s and 40s. Why are some members of the family affected but not the others? Our detective quest takes us through the mechanics of how our eyes respond to light and some rather unexpected side effects of this process.

The setting of the third story moves from the eye to the heart. Here, we are confronted with siblings who suffer heart failure in their 30s. Why does this happen so uncharacteristically early in life? Are the other siblings also at risk? Our search takes us on a tour of the various factors that support the heart's tireless, rhythmic beats: electrical impulses, powerful muscle contractions, and the adhesive structures that hold things firmly in place in the midst of all this energetic movement. In pinning down the root cause, we are confronted with our first taste of genomic uncertainty. How confident are we that our answer is correct?

The challenge of genomic uncertainty is taken to its peak in the fourth story—distraught parents whose two children have passed away of mysterious causes when they were just a year old. Our detective mystery finds itself in uncharted genomic territory as it tackles a character which has never previously been observed. Is this indeed the character responsible for the tragic fate of the two children? While we are still debating this fact, the stakes on the answer are raised considerably because the couple is now expecting their next child. Would this child encounter the same fate?

The fifth story presents a most curious phenomenon—children whose organs are out of place. Our detective quest, after some false starts, hits a real conundrum: the children's genomes appear not to have been inherited from their parents as they should be. How could this happen? Much head-scratching then offers a solution: a rare genomic event that lies concealed under mounds of data.

The sixth story involves a patient whose hemoglobin (the red pigment in our blood) cannot carry and deliver the regular quota of oxygen to various parts of the body. This relatively well-known condition is expected to be a cakewalk for our detectives. Yet, our search draws a blank. This prompts some deeper digging, eventually uncovering the genomic cause where it was least expected, and exposing us to some fascinating genomic jumps in the process.

The seventh story switches gears to cancer. Typically, cancer is a disease of age. One in two or three persons will be diagnosed with cancer at some point in life, more likely in their later years. In contrast, the cancer patient in this story is barely a few years old. How does cancer strike so early? The quest for the genomic cause again draws a blank. Some huddles then suggest a new line of attack. And out pops the answer, but not before it leads us through the delicate balance between division, differentiation, and death that tips a normal cell over to a cancer cell.

The eighth story reminds us that the genome in our cells is constantly under attack, and constantly being defended. The patient in this case has a compromised defense, which leads to both immediate challenges as well as an increased future risk of cancer. The only suspect in our hunt appears to have a strong alibi though. More dogged investigation then breaks this alibi and provides a glimpse into some intricate genomic surgeries that happen in nature.

The final story presents a middle-aged cancer patient and a fundamentally different detective quest. Doctors try many different treatments for the cancer. Each time it bounces back. The goal of the detective quest is to determine the next step in this game of moves and countermoves. Will it succeed?

These are but a few among several stories that have come our way. All are real stories, of real patients whose lives have (often) been impacted gravely due to the genomic quirks described in these stories. These stories only touch upon but do not dwell upon this grave impact; the greater focus remains on the genome and on the detective quest for the genomic cause. Regardless, the gravity of impact is always acting as the backdrop and any attempt at simplification or lightness for lucidity is not intended to disrespect it in any way.

With that introduction, we can launch into our stories. "What!" you might say. No introductory rant on the biology of it all: the genome, chromosomes, DNA, RNA, introns, exons, cells, nuclei, etc.? Indeed, there is much science and engineering that these stories are based on. Each story attempts to connect the world of clinical practice, built upon centuries of careful observation of external form, to the world of molecular biology, with its deep internal secrets. The path between these two worlds passes through the mind-bending world of computer algorithms, which distills large amounts of data down to its essence. The interplay between these three worlds is fascinating, as you will hopefully see. However, the last thing a writer wants is to lose the reader to an extensive overdose of dry facts. Therefore, these facts are woven into the stories themselves in a novel experiment at combining hard science with entertaining storytelling. A *glossary* of terms is at hand though, at the end of the book, just in case you need a place to look things up quickly. On that note, and without further ado, let us dive into our stories of genomic typographical errors and their impact on patient lives.

Story 1

Three Colors or Two?

This story begins when I was 17. I had just finished high school and received college acceptance. The final hurdle was a medical test—usually, a mere formality. The test progressed uneventfully until someone flashed a series of cards and asked me to call out the number written on each card, as in the example below.

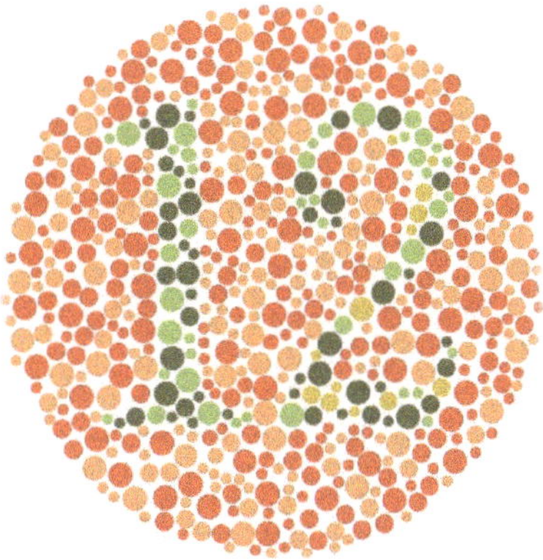

An example Ishihara card. Can you spot the hidden number? Or any number at all? Of course, only if you see this in color and not in black and white.

The candidate who preceded me breezed through his turn without batting an eyelid. My turn came. I saw dots of different colors and different sizes. But these dots did not form a number! At least, not to my eyes. I struggled to find a pattern, but in vain. I muttered out unsure guesses. *Partial color blindness* said my report.

There was a moment of concern, but it allayed soon enough—I was barred from working in mines, but I didn't aspire to do so anyway. What remained was the mystery of this strange disability which had gone unnoticed for the first 17 years of my life. What was it, and how did it come about?

Years later, I was giving a talk on this topic to an audience of about a 100 people. I projected one of these card pictures on screen and asked for a show of hands from those who couldn't spot the number. Five hands went up. It was good to know that I was not alone. More interestingly though, all respondents were males. Why so?

Pinning it Down

The card shown above is an example of the Ishihara test[1], named after its inventor Shinobu Ishihara, a professor at the University of Tokyo, who published this test almost a century ago. As you can see, the card carries dots of various colors. To confound matters further, these dots also have varying sizes, though that is not relevant to our discussion here.

Hidden among these dots is a number formed by dots of one color; we call these dots *number dots.* All other dots, which we shall call *noise dots*, are composed of various other colors and dispersed randomly around the number dots. The color contrast between noise dots and number dots is chosen with some care. For most people, this contrast is prominent enough for number dots to stand out strongly. For some, however, this contrast is just not sufficient. To those eyes, only a mess of dots is visible; the number remains obscure. I certainly fall into this category.

To pinpoint my problem more precisely, I decided to run the following experiment. To keep things simple, this experiment uses a series of Ishihara cards, each comprising just two colors, as illustrated below.

A series of Ishihara cards with the same hidden number, but with increasing amounts of red mixed into the noise dots, from 80 units at the top left to 160 units at the bottom right. At what point can you clearly spot the number? Do note a caveat though: these subtle color differences may not show consistently on all computer screens or in print, and certainly not in black and white.

In these cards, number dots are chosen to be green. Noise

dots also start as pure green, at which point they are completely indistinguishable from number dots. Increasing amounts of red are then mixed into the noise dots in successive cards, Since red and green make yellow, you will find noise dots turn increasingly yellow from top left to bottom right. The hidden number in green reveals itself eventually. Which is the earliest point where it does so is the question.

It is easy enough to create such series on a computer by increasing the amount of red in units of 10. Of course, to guard against our eyes getting trained, the hidden number is chosen at random and the cards are flashed in quick succession. Laptop screens are not suitable for this exercise because colors change quite dramatically when viewed from even slightly different angles. Some tablet screens work well though and that's what I used.

I then called on a few volunteers to participate in this exercise and observed the point when each volunteer could spot the hidden number clearly. When the amount of red mixed was less than or equal to 80 units, most numbers appeared obscure to most volunteers. However, 110 units of added red provided ample contrast for most people to spot the hidden numbers correctly. My eyes were far weaker though. At 110 units of added red, I could barely spot any of the numbers. I could spot numbers effortlessly only when 160 units of red were added. So, I needed 50 extra units of red to perceive this clear contrast. Why?

Going further, was this confusion only between red and green? Would blue be any different? In the blue version of the above experiment, we again start with pure green number dots. We then mix increasing amounts of blue, instead of red, into the noise dots. What happens now? Most volunteers were able to spot the hidden number when 110 units of blue were mixed into the noise dots. My eyes reached 90% success at 110, and 100% at 120 units of blue—only a marginal difference.

In summary, my ability to distinguish blue from green was comparable to most. But my ability to distinguish red from green was significantly poorer than most. There was something clearly wrong.

Or was there?

Sensing Color

Isaac Newton is known for several striking experiments on color. One of these experiments is the starting point in our quest. In this experiment, Newton used a prism to split a beam of sunlight into a spectrum of rainbow colors: violet, indigo, blue, green, yellow, orange, and red. He then experimented with various combinations of these colors.

First, Newton blocked off all the other colors, letting only yellow through. Then he placed a prism in the path of this yellow beam to check if it would split yet again. But the beam just wouldn't split—it stayed yellow.

Next, Newton blocked off all other colors, letting only the red and the green beams through this time. He then combined these red and green beams. To his surprise, he got yellow again, no different from the original yellow beam above. But when he attempted to split this yellow beam using a prism, he was in for an even bigger surprise. This yellow, unlike the previous one, gladly obliged and split into red and green!

Thus, Newton showed that what the eye perceives as yellow is not a single type of light. There are at least two yellows—one that splits into red and green, and another that is pure and doesn't split. However, to the human eye, both yellows appear much the same. How do these very different yellows appear the same to the human eye?

Of course, our eyes have color sensors for detecting colors. We also know that each color has a distinct *wavelength*. As you move away from the violet end of the rainbow toward the red end, the wavelength increases continuously from about 400nm to 650nm (nm stands for *nanometer*, 1 billionth of a meter). Of course, a distinct sensor for each of these wavelengths would be way too many sensors. Instead, the human eye has just three sensors: roughly speaking, a blue sensor, a green sensor, and a red sensor.

With just these three sensors, how does the human eye detect all the colors in the 400 to 650nm spectrum? In particular, how does it detect both types of yellow described above?

As is obvious, red light is detected by the red sensor and green light by the green sensor. When a mixture of red and green light falls on the eye, both sensors detect their respective colors. When both sensors fire together, the brain is tricked into seeing a combination rather than the individual colors. This explains one of the two yellows.

How about the other yellow—Newton's pure unsplittable yellow? The explanation for this is trickier. And the trick is as follows: each of the three sensors responds, not just to a single color, but to a wide range of wavelengths or colors. The picture below should give you a feel for what this means.

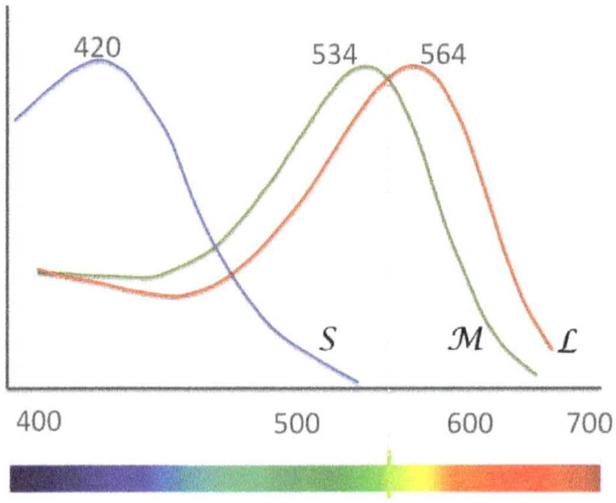

The blue (S), green (M), and red (L) sensor responses to lights of various wavelengths. Just to the right of the dotted vertical line lies the unsplittable yellow.

This picture shows the green sensor responding not solely to green but also to other colors like yellow with nearby wavelengths, albeit less strongly. Similarly, the red sensor responds not just to red but also to other colors like yellow and shades of green with nearby wavelengths, albeit less strongly. If you notice, Newton's pure unsplittable yellow is sandwiched between green and red in the rainbow and has a wavelength that is not too far from either. So, when the eye sees this pure unsplittable yellow, both green and red sensors respond, albeit less strongly. Regardless, both respond with sufficient intensity for the brain to still perceive yellow. This is the explanation for the unsplittable yellow.

Since the three sensors in the picture above are sensitive to a range of wavelengths rather than to a single wavelength, it is better to refer to them as *Short* (*S*), *Medium* (*M*) and *Long* (*L*) wavelength sensors, rather than blue, green and red, respectively. *S* responds the most at 420nm, *M* at 534nm, and *L* at 564nm.[2] These are the three peaks you see in the picture above. In each case, neighboring wavelengths elicit smaller responses, so the hills fade away on either side of their respective peaks.

You would surely have noticed how close the peaks of the *M* and *L* sensors are in the picture above—a gap of just 30nm. In comparison, the peak for the *S* sensor appears much farther away. It is therefore not surprising that green and red are harder to distinguish as compared to green and blue.

However, my ability to distinguish red from green was even poorer than most others. What made my sensors different? Could it be that the gap between my *L* and *M* sensors was even narrower than the usual 30nm? Maybe 20nm? Or 10nm? Or, at an extreme, even 0nm?

At an extreme, if the gap between the two sensors were to become 0nm, there would be no difference at all between the two sensors, and the ability to distinguish red from green would be completely lost. However, I could distinguish red from green quite comfortably—at a traffic light, for instance. So, the peaks of my *L* and *M* sensors certainly did not coincide. However, could they have come somewhat closer? Possibly, along the lines shown in the picture below?

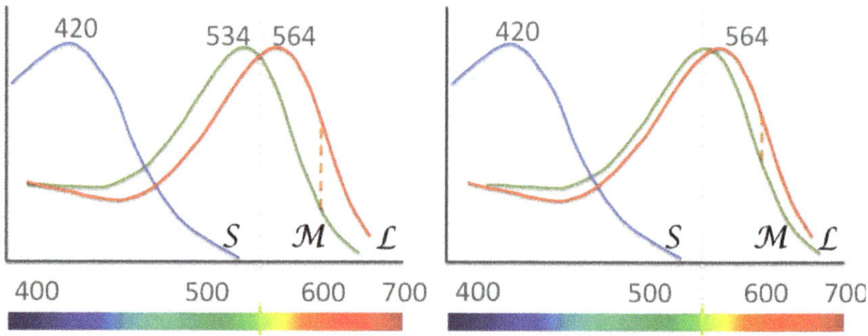

The M sensor response is shifted slightly rightward in the second figure as compared to the normal scenario depicted in the first figure.

In this scenario, where the gap between the *L* and *M* sensors is narrower than usual, the difference in their response becomes even smaller than it usually is. As an example, focus on the dashed vertical line at 600nm in the picture above; it is shorter in the figure on the right as compared to the normal on the left, indicating reduced ability to distinguish red from green.

Was the gap between my *L* and *M* sensors indeed smaller than normal, making my eyes need 50 additional units of red in the Ishihara experiment above? If so, then what could have brought these peaks closer?

Rods and Cones

The answer to this question lies buried deep inside specialized cells in our eyes that are home to these color sensors. These cells are a part of the *retina*, a screen at the back of the eye on which an image of the world in front is projected. Sensors in these cells detect this image, and nerves then carry the detected signal to the brain.

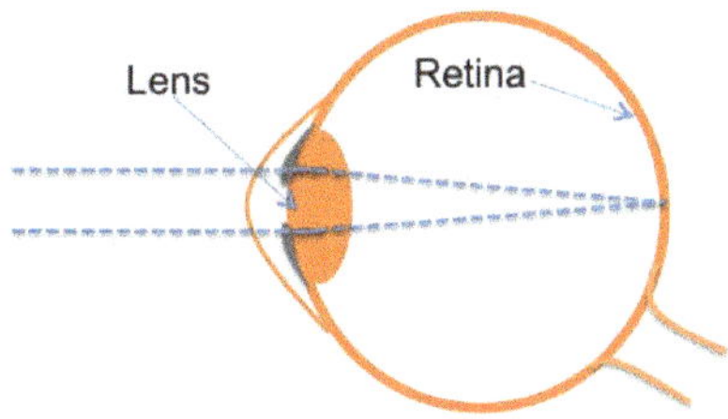

Light focused on the retina by the lens in the eye.

There are a few different types of light-sensing cells in the retina. The most prevalent cells, called *rods*, are very light-sensitive, allowing us to see somewhat even at night or in the dark. They are not color-sensitive, however, and thus not of our interest in this story.

Less prevalent are cells called *cones*, which appear mostly at the center of the retina. Color sensors are present in these cones. Their concentration at the center of the retina allows us to see color nuances better where our eyes focus rather than at the periphery of our field of vision. But having fewer cones than rods restricts our ability to perceive color at night. Here is how rods and cones might be distributed in the retina.

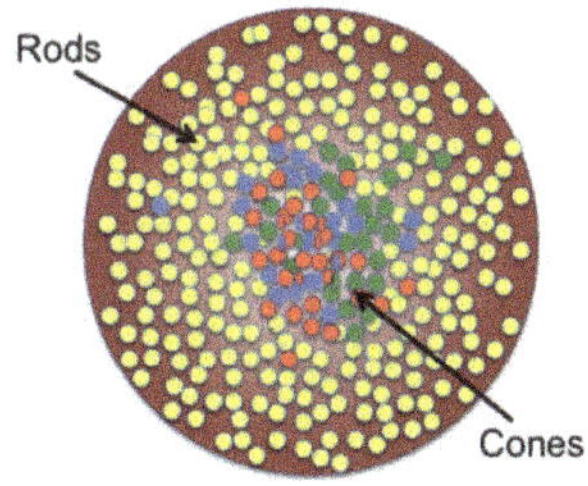

A close-up of the retina, showing rods and cones.

To pursue the answer to our mystery, we zoom into one of these cones. The picture below shows a cone in profile, as also a rod, just so you see how these names came about.

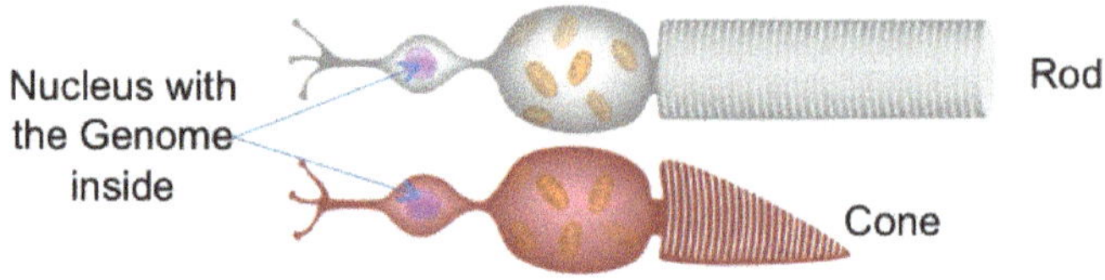

A close-up of the rods and cones themselves; rods are cylindrical as shown, cones taper off to a point.

Continuing our journey inward, note the various *organelles* that appear inside a cone. These are to a cell, what organs like the heart, brain, and kidney are to the body. One such organelle is the *nucleus*, marked in the picture above. Inside this nucleus lies the *genome*—the storehouse of recipes for the creation of many molecules, including, of course, the color sensors we met earlier. The quest for a solution to my red-green mystery takes us into this storehouse next.

The Genome

Of course, the genome is tiny in size, so it is not visible to the eye even through microscopes. We can imagine though that the genome is a collection of books.

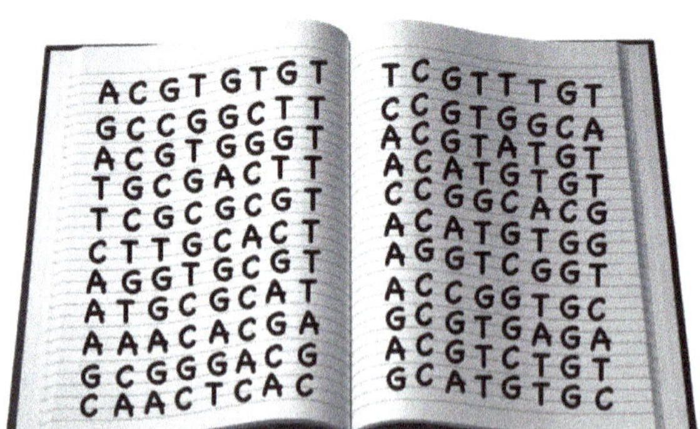

One of the genome books.

An entire copy of this book collection is carried by each and every cell in our body (barring a few exceptions, such as red blood cells, which we will meet in a later story). Unlike books written in English, which use the 26 characters of the alphabet along with various punctuation marks, the alphabet in these books comprises just 4 characters: A, C, G and T. The total number of characters though is a mind-boggling six billion!

Each book in the genome is called a *chromosome*. There are 46 such chromosomes, organized into 23 pairs. The two chromosomes in each pair carry very similar (but not identical) character sequences. Here is how we can imagine all these chromosomes, much like books on a bookshelf so only the book spines are visible.

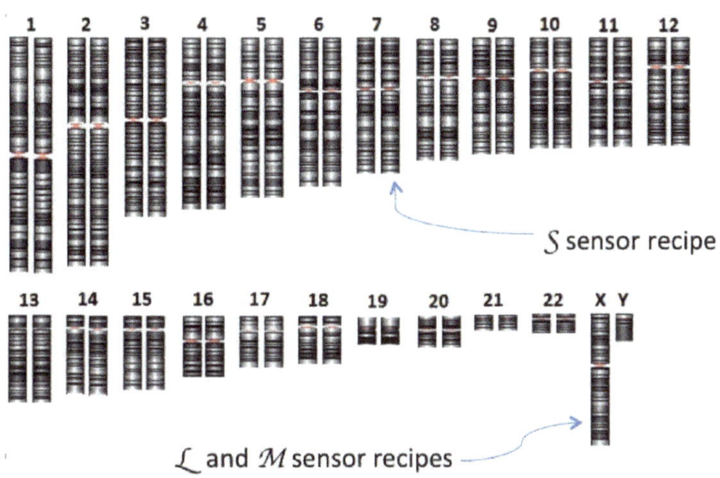

23 pairs of books, i.e., chromosomes, as if on a bookshelf with their spines visible. These books have different sizes and are therefore shown with different heights. The 23rd pair indicates this is a male subject. The recipe locations for the three color sensors are shown as well.

In the picture above, you can spot one exception: in males, the

23rd pair comprises two very different books, labeled X and Y. In contrast, females have two very similar copies of the X book, and no Y at all. Since X and Y chromosomes determine the sex of a person, they are called sex-chromosomes.

How do these books find their way into each and every one of our cells? It starts with our parents, who are already that way. Each parent contributes a single cell—a sperm cell or an egg cell, as the case may be. This special cell has just 23 chromosomes, and not 23 pairs of chromosomes. When the sperm and the egg fuse, each contributes 23 chromosomes, and the resulting cell now gets 23 pairs of chromosomes. This cell then divides repeatedly to generate the tens of trillions of cells that constitute an individual. With each division, the books are copied so daughter cells get a complete and almost faithful copy.

Each cell in our body thus inherits six billion characters—surely a lot of information. Buried in this ocean are stretches of text that describe recipes for the manufacture of color sensors and of many other useful molecules. These special recipe stretches are called *genes*. Molecules manufactured from these recipes are called *proteins*. Color sensors too are a type of protein.

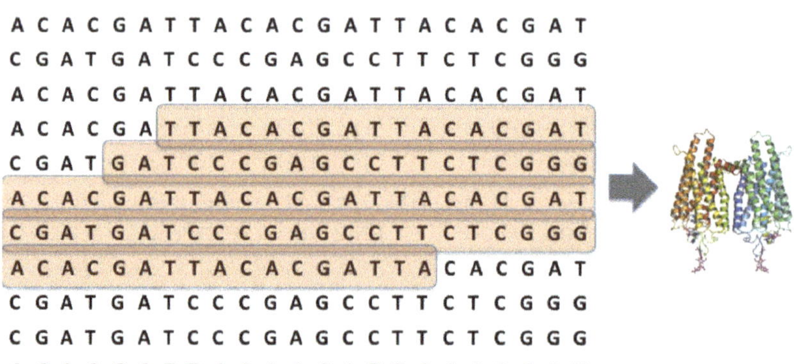

A gene, and the protein created using this gene's recipe.

There are about 20,000 genes in our genome. Each gene has its own name and carries recipes for its own distinctive collection of proteins. Indeed, we will meet several genes by name in these stories.

Interestingly, even though almost every cell in our body contains the genome in its entirety, each cell picks and chooses which of the 20,000 recipes to actually execute. Heart cells pick recipes that create proteins which sustain the heartbeat. Muscle cells pick recipes that create proteins which facilitate expansion and contraction. And cone cells in the retina pick recipes which create color sensors. For the mystery at hand, these are the genes of our interest. Do these recipes hold a clue to my red-green confusion?

Color Sensor Recipes

Remember we have three distinct types of color sensors: *S*, *M* and *L*. It turns out that cone cells are a picky lot—each cone picks only one of these three sensor recipes to execute. Accordingly, there are three distinct types of cones. The first type manufactures the *S* sensor from the recipe described in a gene called *OPN1SW* (or just the *S gene* for short). Meanwhile, the second manufactures the *M* sensor from a gene called *OPN1MW* (or just the *M gene*), and the third manufactures the *L* sensor from a gene called *OPN1LW* (or the *L gene*).

The *S* gene is present on chromosome 7 in the genome. In contrast, the *L* and *M* genes are both located on the X chromosome, which you might recall is one of the two sex-determining chromosomes. Since females have two copies of chromosome X while males have only one, this makes for some interesting gender variations in color perception, as we shall see later. For the moment, we examine the recipes in the *S*, *L* and *M* genes to look for clues to my strange condition.

Recipes, Interrupted

Rather inconveniently, gene recipes are not written out as consecutive characters in the genome. Instead, they are written out in blocks of characters, called *exons*, with intervening stretches of characters called *introns.* Only exons carry the recipe, introns don't. For instance, the recipe might be composed of just the underlined characters in this text: AGCGTAAGGGCGTAGAGGT. This is just an illustrative example. In reality, exons carry many tens or few hundreds of characters, and introns may be even longer. Our cells know how to skip over these long introns when reading and executing the recipe. Here is how we can represent the exons and introns for the S gene recipe graphically.

Exons (rectangles) and introns (horizontal lines) in the S gene.

This recipe is written out in five exons, with the text in the four intervening introns being skipped when this recipe is executed. Here is the corresponding picture for the L and M genes.

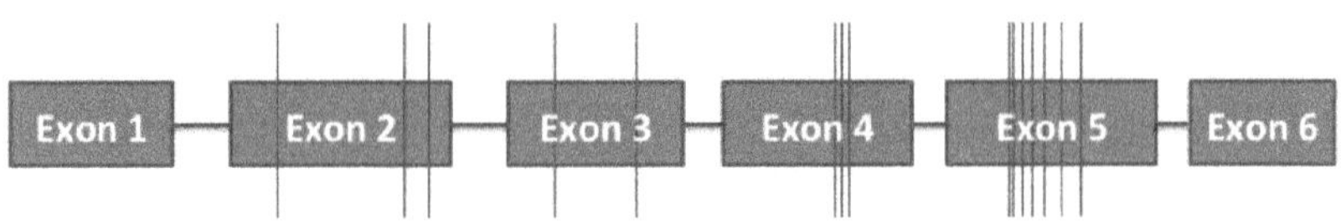

Exons and introns for both the L and the M genes. The dark vertical lines indicate 15 places where the two genes differ.

This recipe is written out in six exons instead of five. However, note that we have shown only one picture above, despite having two genes: L and M. These two genes are so uncannily similar that we are indeed able to depict both in the same picture. Both genes have the same number of exons. The recipes written in these exons are also almost identical. Almost, that is—there are 15 characters at which the two recipes differ, and these locations are marked in the picture above by tall vertical lines. The two resulting proteins, the L sensor and the M sensor, respectively, are therefore very similar as well. That is why their corresponding response curves are very close to each other. In contrast, the *S* gene is very different, with many more differences from either the *L* gene or the *M* gene.

These 15 character differences between *L* and *M* account for the 30nm difference in response between the *L* and *M* sensors—534nm for *M* vs 564nm for *L*. Of these, just two differences in exon 5 account for as much as 21nm,[3] a fact that will turn out to be instrumental in due course. One difference in exon 3 accounts for another 5nm. The rest is spread over the remaining 12 differences.

Though small, this difference of 15 characters between the *L* and *M* sensors suffices to create sufficient distinction between red and green in most people. What made me different then? Was the difference between my personal *L* and *M* recipes even smaller than 15 characters? Could that explain my increased red-green confusion?

My Recipes and Yours

It should come as no surprise that each individual has his or her own distinctive genome. Indeed, this is what makes each of us look and feel different from the other. What might be surprising is how few characters really differentiate one person from another—just one in 1000 characters, or 0.1%, roughly speaking. At this rate, any two of us will differ in about five-six million characters of the six billion in our genomes. These inter-person variations imply that not everyone may have a 15-character difference between their *L* and *M* recipes. Some individuals may have fewer than 15 differences, and some may have more.

For instance, I know I have a *variant* at a particular character in exon 3 in my *M* gene; what is an A for most people is a G for me. It is very close to a key character that causes a 5nm difference between the *L* and *M* sensors. Could this variant, shown below, be the cause of my color perception problems?

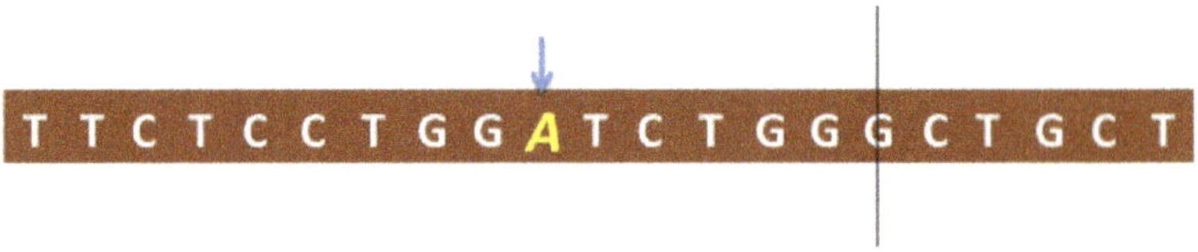

A close-up of exon 3 of the M gene. The A highlighted is where I differ—most people have an A while I have a G here instead. The tall vertical line depicts one of only two characters in exon 3 that are different between the L and M genes, in most people. This character contributes 5nm of the 30nm difference between the M and L sensors.

The answer, unfortunately, is no. This A to G variant is unlikely to be the cause of my problem because it is a very *common variant*—about 30% of the X chromosomes in all the people on earth appear to carry a G, while the remaining 70% carry an A. If this variant could indeed increase red-green confusion then many more than five in 100 people would fail to recognize numbers in Ishihara cards. The cause must lie elsewhere. We need to look further.

Cooking Up New Recipes

So, of course, your recipes are different from mine, and that makes for great variety. Interestingly, nature has tricks up its sleeve to continually generate even greater variety from this variety. One such trick is of great relevance here, but needs some context.

Recall that the genome comprises 23 pairs of books called chromosomes. The 23rd of these pairs consists of the sex-determining X and Y chromosomes. Males have an X and a Y, while females have two X chromosomes.

Now consider a mother and her two X chromosomes. An egg cell created by this mother carries only one X chromosome. A question poses itself now: which of her two X chromosomes makes its way into this egg cell? This is an important question, because this X chromosome is inherited by her child. If the mother's two X chromosomes were identical then the answer wouldn't matter. But that is not the case; her two X chromosomes are very similar but not identical—they do carry differences. Some of these differences could have a drastic impact on life if inherited, as we will see in a subsequent story. Therefore, the answer to the above question assumes significance.

The answer is also truly surprising. The X chromosome that finds its way into the egg cell and is eventually inherited by her child is neither of her two X chromosomes. Rather, it is a hybrid of her two X chromosomes, created by a process called *crossing-over.*

During crossing-over, each chromosome is divided into sections. Then odd-numbered sections from one chromosome are stitched together with even-numbered sections from the other to create a mosaic-like hybrid. In doing so, the egg-creation process cooks up a whole new combination of recipes from two existing sets of recipes. This continuous experimentation with new combinations generates the variety that we see all around us. Here is an illustration of how the hybrid is formed.

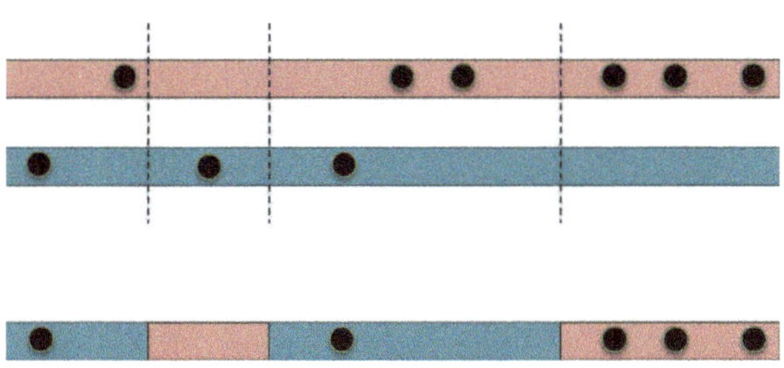

The mother's two X chromosomes on top. The dots indicate character differences between these two chromosomes. At the bottom is the hybrid chromosome.

Of course, breaking things apart and then putting them back together always leaves scope for error. Could the cause of my condition also lie in some such typographical error made during crossing-over? The answer relies on some more insights into the crossing-over process.

The dotted vertical lines in the picture above, where sections switch, are called *cross-over breakpoints*. An important property of these breakpoints is that each such breakpoint cuts through the same location in the two chromosomes. In particular, the situation shown in the picture below rarely happens.

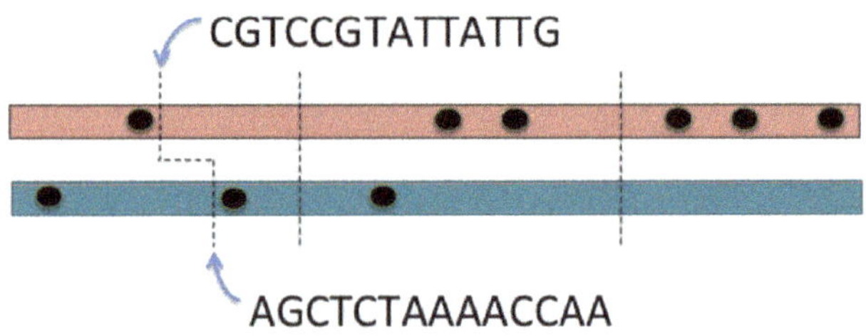

An unlikely scenario: three cross-over breakpoints, of which the first cuts through the two chromosomes at different locations. The text around this breakpoint in the two chromosomes is also shown.

Here, the first breakpoint is shown cutting through different locations in the two chromosomes. If this sounds confusing, let us go back to the analogy where each chromosome is a book. In this analogy, the picture above is tantamount to tearing off the two books at different page numbers. And that rarely happens; the two books shown in the picture above must be torn off at the exact same page numbers during crossing-over.

Of course, there are no page numbers in the genome. Instead, nature has to use a surrogate for these page numbers. That surrogate is simply the content on these pages. If the text content looks similar, the guess is that the two books have indeed been opened to the same page and crossing-over can proceed. Since the text content depicted in the picture above is very different in the two chromosomes, the lopsided cut shown in that picture is avoided.

So, that is how a mother's X chromosomes create the hybrid chromosome which her child inherits. Of interest to our story here is the effect this process has on the *L* and *M* genes. Both genes are on the X chromosome. In fact, they are right next to each other on this chromosome. Even more interestingly, in what makes for a very odd arrangement, there are two copies of the *M* gene right next to each other, as shown below.

The L and two M genes next to each other on the X chromosome.

The recipe in only the first of the two *M* genes is used for construction of the *M* sensor. The second gene is probably not used,[4] but is nevertheless present in the genome. This rather odd arrangement is a possible source of complication, as we will see soon.

Now imagine what happens when a cross-over breakpoint cuts through these genes. This breakpoint along with the resulting hybrid chromosome is depicted in the picture below.

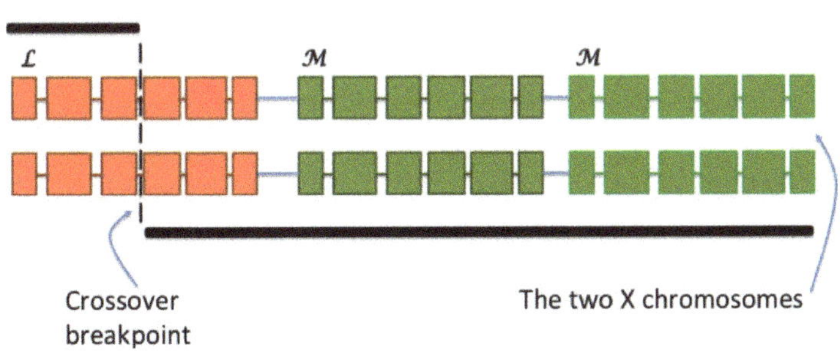

The mother's two X chromosomes, zoomed into the region with the L and M genes. The hybrid created is indicated by the thick, dark bars.

How does this hybrid, obtained by joining the two dark bars in this picture, differ from the two original chromosomes? Since the two chromosomes have very similar character sequences to begin with, these hybrid genes are not hugely different from the original. Therefore, the L and M recipes remain largely uncompromised in this hybrid.

Well, then why make such a fuss about hybrid chromosomes and cross-over breakpoints? Bear with me a little longer, for another question poses itself at this point. What if the cross-over breakpoint were to cut through different locations in the two chromosomes? For instance, imagine the breakpoint depicted in the picture below.

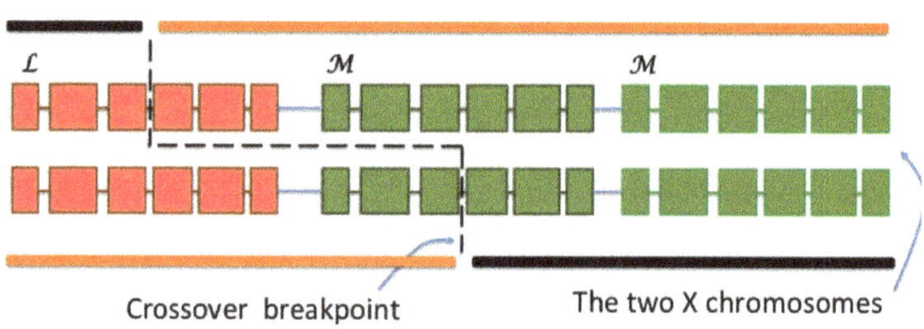

A lopsided cross-over breakpoint and the two possible hybrid forms, indicated by thick bars of different colors.

Wait a minute! Can this happen at all? We ruled out such lopsided breakpoints earlier because crossing-over requires that the character sequences around a breakpoint be very similar in the two chromosomes. However, we overlooked an important fact in the process—the L and the M genes are very similar to each other. Therefore, in this special case, the sequence similarity condition is indeed satisfied by the lopsided breakpoint shown in the picture above.[2] This lopsided breakpoint is thus blessed, or at least allowed to function, by sheer accident.

What is the hybrid created by this lopsided breakpoint? Based on the thick bars in the previous picture, there are two possible combinations, as shown below.

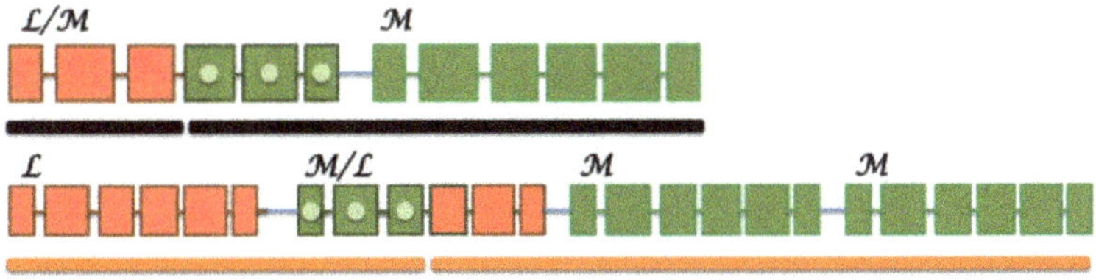

Two possible hybrid forms that the lopsided cross-over breakpoint can lead to. For black-and-white readers, dots in the hybrid genes indicate which exons come from M.

In the first case, there is a hybrid L/M gene followed by a single intact M gene; so just two genes in all instead of three. In the second case, there is an intact L gene followed by a hybrid M/L gene followed by two M genes; so four genes in all instead of the customary three. In either case, the gene for one sensor is intact and the gene for the other sensor becomes a hybrid of the L and M genes—a serious change in the recipe for that sensor.

If you haven't already noticed, here is an interesting fact: exon 5 is green in *both* genes in the first hybrid form, and red in *both* genes in the second hybrid form (recall, only the first two genes are relevant in the latter). Usually, exon 5 should be red in the first gene, and green in the second. Not so in these hybrid forms. And this has an interesting consequence.

Recall that differences in exon 5 account for 21nm of the 30nm difference between the M and L sensors.[3] This 21nm difference is absent in these hybrid forms. Therefore, the gap between the two sensors in these forms is not more than 30-21=9nm. So, a lopsided cross-over breakpoint can indeed bring the two sensors much closer to each other than they usually are.

This suggests that a lopsided cross-over breakpoint in one of my female ancestors generated one of these hybrid chromosomes, which she passed down into an egg cell she created, whence it passed on to her child. Her child might in turn have passed this hybrid chromosome on to his or her children, and so on. Over time, this strange new chromosome X may have found its way into me. With one of its sensor genes being replaced by a hybrid, this chromosome was possibly the genesis of my color perception problem. How could I confirm this hypothesis?

A Peek at the Recipes

The only way to verify whether I have pure L and M genes or a hybrid gene is to peek into my genome and see what it holds. Of course, the genome is tiny; the naked eye can't see it, and neither can a microscope. Cleverer methods are needed.

There are tiny, naturally occurring molecules whose daily job it is to read the genome, execute the recipes coded in various genes, make proteins from these recipes, and make entire copies of the genome as cells divide. Scientists have figured out how these tiny molecules can be manipulated in clever ways so they actually provide a read-out of the genome on to a computer disk. This process is called *sequencing* of the genome. It works as follows.

We start with a little saliva collected in a tube (or blood, or any other body tissue, for that matter). This saliva contains cells from which my genome is extracted using a chemical process and subjected to sequencing using a process called *NGS*, short for *Next-Generation Sequencing*—a term used to contrast modern genome-sequencing methods from previous, far less powerful, and far more expensive methods.

Unfortunately, the NGS process is not powerful enough to read each of the chromosomes in my genome end-to-end in a single step. Instead, it requires that we proceed in a much more convoluted way, as follows. We chop these chromosomes randomly into small pieces, or *reads*, of about 100 characters each. Tiny genome copying molecules are then let loose on these reads. These molecules make copies of each read, one character at a time. With each character they copy, they are tricked into emitting a visual cue; the color of this cue depends upon the character copied. These color cues are captured by a camera and then processed by a computer to extract the character sequence for each read. The contents of each read are now available. Unfortunately, the order of reads in the chromosome has been completely lost.

In other words, we've managed to tear our genomic books into shreds of paper. We know what is written on each piece of paper. We now need to put these pieces of paper together in the right order. This requires solving a massive jigsaw puzzle, which is no mean task.

Solving the Jigsaw Puzzle

A monumental effort to put these pieces together was launched in the 1990s; it was called the *Human Genome Project.*

The focus of this project was not the genome of a single individual; rather, it was the pooled genome of a few unnamed and presumably healthy individuals. The effort used huge amounts of computing power as well as several follow-up experiments to solve this massive jigsaw puzzle. Eventually, it culminated in the publication of the first-ever almost-complete human-genome character sequence in 2004.[5] This sequence has a name—the *reference genome sequence*, or just the *reference sequence* for short.

The total cost of this pioneering endeavor is estimated to have been a few billion dollars! If that boggles your mind, then consider the fact that returns from this investment by way of applications enabled are estimated to run into hundreds of billions of dollars. Indeed, among other things, this reference sequence turns out to be a blessing when it comes to solving the jigsaw puzzle for my genome.

The reason the reference sequence helps is that the genome sequences of any two individuals are very similar—differences occur on average at the rate of just one in 1000 characters. This means my genome sequence and the reference sequence are very similar. Therefore, a read from my genome, e.g., AGGTTCTG, is likely to be present somewhere in the reference sequence as well, in slightly altered form if not in identical form. An example with a single alteration appears next.

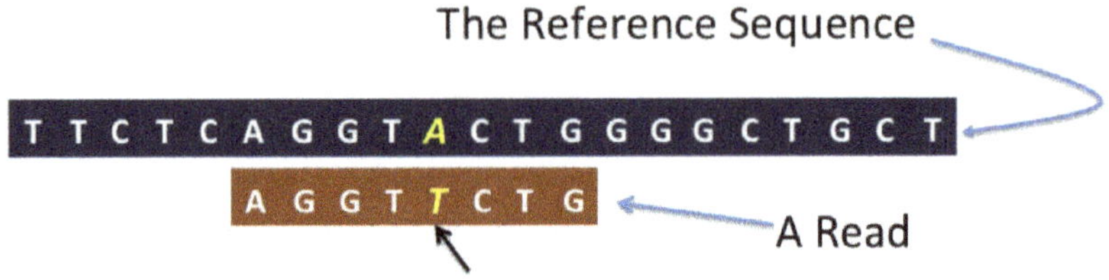

A read from my genome anchored to the right place in the reference sequence. This is a match with one character altered.

The similarity between my genome and the reference sequence suggests a procedure by which we can solve our jigsaw puzzle without much sweat. We take each read in turn from my genome and search for a match in the reference sequence. This search keeps an eye out for both perfect matches as well as matches with slight alterations. Given the billions of characters in the reference sequence, this procedure is too laborious to run by hand, but eminently within the reach of a computer. This so-called *read alignment* algorithm attempts to anchor each read to its rightful place in the reference sequence. Once that is done, the jigsaw puzzle is effectively solved. This procedure is fully automated and can be accomplished quickly on a powerful computer.

There is a small, added complication though, but one that is helpful in many ways. It is difficult to isolate just one cell in the body and sequence the genome in that one cell alone. Instead, we take several cells together, chop up the genomes in all these cells into reads, obtain the sequence of each read, and finally anchor each read to its rightful place in the reference sequence using the read alignment algorithm. We now have many different reads anchored at every location in the reference sequence. Why this is helpful, we shall see shortly, as we get to the final step of unraveling the mystery of my red-green color confusion.

So, that's the crux of how genome sequencing works. An elaborate procedure, no doubt, but a powerful one at that—it gives us much of my genomic sequence at modest cost. Sequencing the entire genome with this procedure costs just a few thousand dollars at the time of this writing. The cost drops to a few hundred dollars if we chemically extract out just the exons of the various genes from the genome, and sequence only these. Since gene recipes are carried by exons, sequencing only the exons suffices for many purposes. Indeed, most stories in this book use this approach, and so does this one.

Coming back to our mystery: what do the reads from my genome, now aligned to the reference sequence, tell us about my L and M genes?

Unraveling the Mystery

We examine the reads that the read alignment algorithm places under the exons of the L and M genes. Here is the picture. A keen eye may just be able to spot a clue to my color confusion mystery here. If not, simply read on.

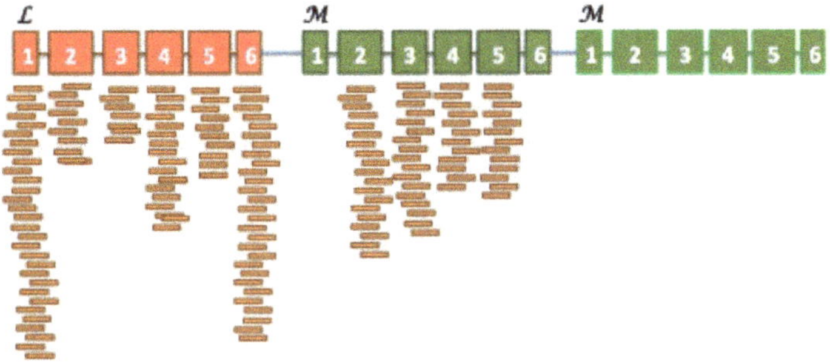

Reads from my genome piled up below the exons of the L and M genes in the reference sequence.

A few oddities stand out at first glance. We see several reads placed below the exons of the first of the two M genes. However, the second M gene has no reads at all. A moment of thought explains why: the two M genes have identical character sequences in the reference sequence. So, any read which finds a match in the second

M gene will also find a match in the first *M* gene. The read alignment algorithm then places all such reads below the first *M* gene as a matter of convention.

For much the same reason, we see no reads below the first and the last exons of the *M* gene. These exons are identical in the *L* and *M* genes. The read alignment procedure, lacking any means of disambiguation whatsoever, simply places all such reads below the *L* gene.

With these distractions behind us, we now focus on the crux—the four central exons: exons 2, 3, 4 and 5. Each such exon in the *L* gene has at least a couple of character differences from the corresponding exon in the *M* gene. Consequently, the read alignment algorithm has no trouble discriminating between the *L* and *M* versions of these exons, and thus places reads accurately below one or the other.

We step through each of these four exons in turn, counting the number of reads on each. Since there are two instances of *M* in the reference sequence but only one instance of *L*, we expect half as many reads in *L* as compared to *M* for these exons. We can depict this expected pattern as **50%, 50%, 50%, 50%**. In other words, we expect the number of reads on exons 2, 3, 4 and 5 of the *L* gene to each be 50% of the corresponding number on the *M* gene.

Note, **50%, 50%, 50%, 50%** is what we expect. But when we do the actual counts for my genome, what do we get? We get a rather unexpected pattern: **31%, 23%, 112%, 75%**.

Isn't this a strange pattern? Exons 2 and 3 have low numbers, while exons 4 and 5 step up substantially. What we expected were four roughly equal numbers, not two of one kind and two that are much higher. How do we explain these strange numbers?

Maybe the hybrid forms we encountered earlier hold a clue. Remember those hybrid forms generated by lopsided cross-over breakpoints? Could these strange numbers arise because I had inherited one of these hybrid forms?

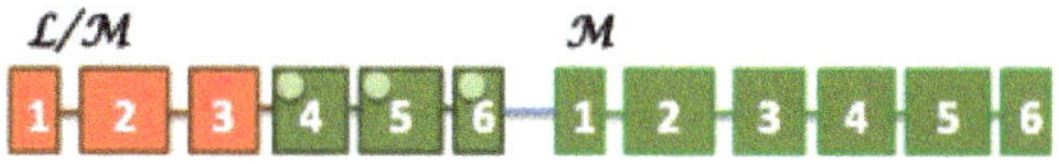

The first hybrid form. For black-and-white readers, dots in the hybrid L/M gene indicate exons from M.

Take the first hybrid form, reproduced above so you don't need to go back several pages. This is the unusual form with just two genes. We count how many instances we have for each of the exons. For exons 2 and 3, we have one instance each of L and M. So, on these exons, we expect as many reads in L as in M—in other words, a **100%, 100%** pattern. In contrast, for exons 4 and 5, we have no instances of L at all. So, we expect no reads in L, and therefore a **0%, 0%** pattern. If indeed I had this hybrid, then the combined pattern must be **100%, 100%, 0%, 0%**. Unfortunately, this is nowhere close to my actual numbers: **31%, 23%, 112%, 75%**.

With bated breath, we move on to the second hybrid form below, the one with four genes. Will we have better luck this time?

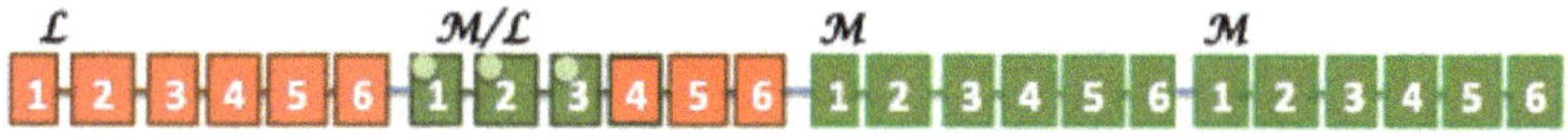

The second hybrid form. For black-and-white readers, dots in the hybrid M/L gene indicate exons from M.

For exons 2 and 3, we have one instance of L and three of M now. That should give us a **33%, 33%** pattern. In contrast, for exons 4 and 5, we have two instances each of L and M. That should yield a **100%, 100%** pattern. If I had this hybrid, then the combined pattern must be **33%, 33%, 100%, 100%**. Two small numbers and then two much bigger ones. Doesn't that sound familiar?

My actual pattern **31%, 23%, 112%, 75%** is not terribly far off. It is not a perfect match, but that is not surprising. After all, if you toss a fair coin ten times, you don't always get exactly five heads; there is some natural variation. It is close enough though. Close enough to suggest that my genome carries the second hybrid form shown in the previous picture, with a true red gene and a hybrid green gene. Voila!

For confirmation, we attempt to verify the presence of this hybrid gene in my genome using a completely independent method—the classical method used to sequence genomes, a gene at a time, before NGS established itself as the method of choice. The result puts it beyond doubt: indeed, my genome carries this second hybrid.

In summary, my L gene is intact. But my M gene has been replaced with a hybrid M/L gene, and accordingly, my M sensor behaves much more like my L sensor than it should, reducing the 30nm gap between the two to below 9nm. This markedly diminishes my ability to distinguish red from green. And that makes me a complete failure at spotting numbers in Ishihara cards, even though I have no trouble at all at a traffic light. All due to an accidental cut-and-paste error in my genome. My S sensor is normal though, so my ability to distinguish blue from green remains normal.

Wrapping Up

Three decades have elapsed since I became aware of my partial color blindness and before I could finally understand its cause—an accidental cut-and-paste error in the genome. I now proudly wear the *Deuteranomaly* label, meaning my M sensor behaves much more like my L sensor than it normally should.

Things could have been worse had I inherited an X chromosome obtained via an even more lopsided crossing-over event, as shown in this picture.

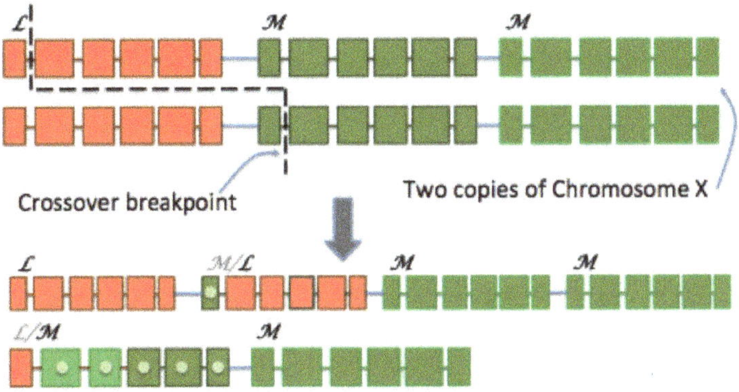

Another interesting crossing-over scenario that makes both sensors identical.

My functional *M* gene would have been completely replaced by an *L* gene, or vice versa, making the *M* and *L* sensors identical (keep in mind that exon 1 is identical in the two genes, so that difference doesn't matter). That would leave me with only two sensors instead of three, and the ability to differentiate between red and green would be totally gone.

The famous scientist John Dalton noticed his own inability to distinguish between colors in the red-green spectrum almost two centuries ago. It was only in 1995 that sequencing of his *L* and *M* genes from preserved eye remains[6] confirmed that he had only the *L* gene, and lacked the *M* gene altogether.

Interestingly, most deuteranomalous individuals are male. Why? The *L* and *M* genes appear on the X chromosome. Females have two versions of this chromosome, so sensor recipes in both these chromosomes will need to be mutated for deuteranomaly to arise in a female. Males, because they carry a single X chromosome, are much more susceptible.

The similarity between the *L* and *M* genes, their nearness on the X chromosome, and the relative distinctiveness of the *S* gene on chromosome 7, makes confusion between *L* and *M* much more likely than between L and S, or between M and S. Therefore, confusion between red and green is much more prevalent than between red and blue, or between green and blue.

Both my children can read Ishihara cards effortlessly. When I struggle on these cards, they think I am playing dumb. They can't believe that is the way I am made. If I were a child, I can well imagine parents and teachers pushing me to practice harder and harder on these cards; little would they know that no amount of practice helps. Genomic characters are of that nature sometimes—discreet, yet overwhelming.

One could argue that the M/L hybrid gene is fairly innocuous. Maybe so, but that is not the case with several other character variations in the genome. We will later see examples of genetic variations that are far more serious—variations that could debilitate or even kill. We will also see examples of how mankind can navigate around these variations or confront them.

Story 2

The Picture Gets Blurred

Let us now switch to the story of a patient, whom we shall call *Dia*. *Dia* had normal vision into her late 30s. Then, gradually, the picture began to blur. While reading, characters began to lose clarity. When in conversation with someone, the person's face appeared unfocused. This blurriness worsened over the years. A blank patch at the center of the vision field became a permanent fixture by the time she was in her late 40s. Faces became hard to recognize. Reading was barely manageable with the aid of a magnifying glass. Multiple doctor visits failed to yield a conclusive diagnosis, let alone offer a cure. I imagine this is how a face would appear to her eyes.

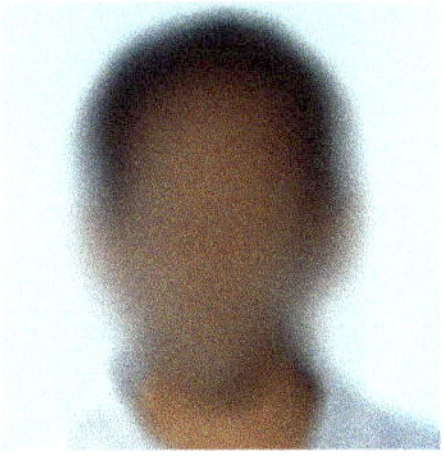

Blurring of central vision.

Dia was not alone in her ordeal. She had seven siblings: four brothers and three sisters. One brother *B* and one sister *S* shared this affliction; the other siblings remained unaffected. *B*'s active career in the police had been somewhat limited by this impediment.

Several attempts were made at seeking mainstream and alternative medical help but none could arrest the progression of his vision loss. Now in his 70s, he could neither read nor watch TV, could barely recognize faces, and could navigate outdoors only with great difficulty. Surprisingly, *Dia*'s mother *M* had lived well into her 80s with perfect vision. *Dia*'s father *F* had passed away in his early 30s; whether he would have suffered from vision loss if he had lived longer will never be known. Here is a picture showing *Dia*'s family.

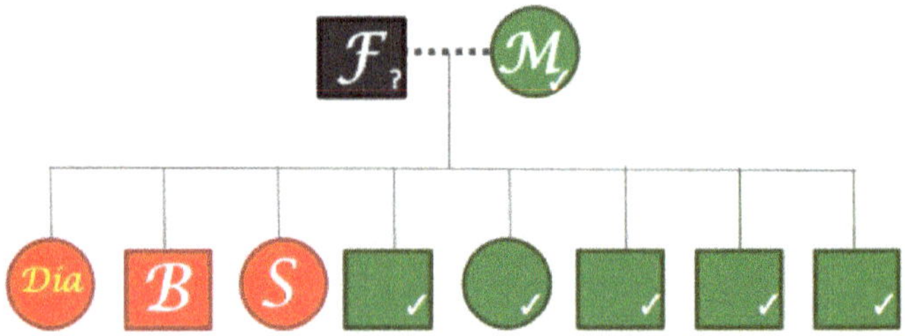

Dia's family: her mother M and father F had eight children. Three of these children are affected: Dia, one brother B, and one sister S. The others, marked with a tick, are not. The question mark for F indicates uncertainty on account of early death. Note that females are depicted as circles and males as rectangles.

Dia and her siblings, *B* and *S*, had adapted to live fruitful lives in spite of this diminishing faculty. The frustration of several unanswered questions remained though. Why was this disorder so common in the family? Whom did it choose to target and whom did it choose to spare? Why did it bide its time and set in only in the third or fourth decade of life?

There was another serious question as well: would this phenomenon resurface in the next generation? *Dia* and her siblings had several children. *B*'s three daughters ranged in age from the early to the late 40s. The second daughter had started showing signs of central vision loss; the other two daughters were unaffected as yet. *Dia*'s children included a daughter in the early 40s and a son in the late 30s; both were unaffected. *Dia*'s children and *B*'s children are shown in the picture below.

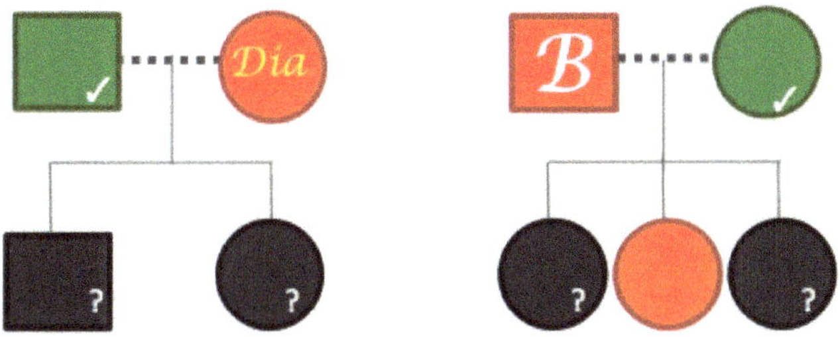

Dia's spouse and children, and B's spouse and children. Individuals in gray with a question mark could yet go on to develop the disease.

Would the next generation remain unaffected, or was it just a matter of time? Going further, was there hope of a cure on the horizon?

Pea Plants and Hidden Characters

Our search for an answer begins with the following thought: since many individuals in *Dia*'s family faced vision loss, an offending character in the *genome* is the likely cause. Remember, the genome is a collection of books comprising six billion characters, copies of which are found in most of our cells. No two individuals have the same sequence of characters, unless they are identical twins; there are always differences.

So, imagine this hypothetical offending character, present in the genomes of individuals affected with vision loss, but absent from the genomes of others. For instance, affected individuals might have, say an A, while all others might have a G. Our goal is to identify this offending character by scanning the genomes of *Dia* and possibly her affected siblings. We have to be careful in this search though, for Mendel's famous pea plant experiments suggest that this offending character need not always announce its presence quite so clearly.

Mendel carefully bred and studied as many as 29,000 pea plants between 1856 and 1863, and observed them for various traits: flower color, plant height etc. One of the traits he studied was seed color, which was either green or yellow.

To study how this color was inherited, Mendel started with pure-bred green and yellow plants, obtained by repeatedly breeding plants of that seed color with each other. He then experimented with breeding across colors, and this offered some rather unexpected observations.

When a plant with green seeds was bred with another with yellow seeds, all the offspring plants produced only yellow seeds. It wasn't as if some plants produced yellow seeds and some produced green seeds, or as if each plant produced some yellow and some green seeds. Every first-generation offspring plant produced yellow seeds only! The color yellow seemed to *dominate* the color green completely, obliterating it in the process. Indeed, it seemed as if the color green had been forgotten altogether.

Accordingly, when these first-generation offspring were bred amongst each other, Mendel expected the second-generation offspring to also produce only yellow seeds. Instead, to his great surprise, the color green made a comeback—a quarter of the second-generation offspring produced green seeds.

Clearly, the first-generation offspring, while producing only yellow seeds, had retained some secret memory of the color green. And this secret memory was brought to the fore in the second generation. How did this happen?

We now have a simple explanation for this seemingly strange hide-and-seek behavior where one version of a trait suppresses the other in one generation but lets it re-emerge subsequently. The source of this secret memory is indeed the genome. The decision to create yellow seeds versus green seeds is determined by one of the characters in the genome. The trick lies in the fact that the 46 chromosomes which constitute the genome come in pairs. The two members in a pair are inherited one from each parent, and are very similar to each other, but not identical—on occasion, one member of the pair might have an A while the other might have a different variant, say G. The interaction between these two versions of the same character produces the observed effect. Read on to see how.

Focus on the character that drives the yellow–green decision. A variant of this character, say A, might produce yellow seeds, while another variant, say G, might produce green seeds. This yields three possibilities for the chromosome pair that carries this character: both members of the pair may have a G (call this **GG**), or both may have an A (call this **AA**), or there might be one of each (call this **AG**).

Mendel started with pure-bred yellow and green plants—**AA** and **GG**, respectively. Then, as generations followed, these characters mixed in various ways illustrated here.

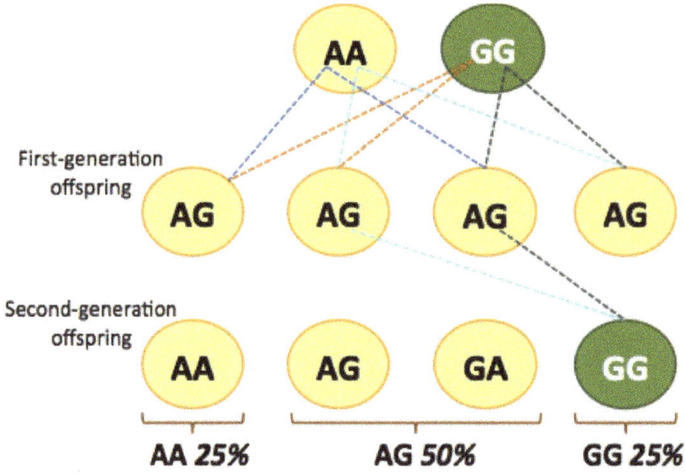

A pair of genomic characters in the parents and in two generations of offspring.

The first-generation offspring were all cross-bred and inherited one version of the character from each of their pure-bred parents. In other words, they were all **AG**. It turns out that the variant A *dominated* variant G, forcing its yellow over G's green. Therefore, all first-generation offspring produced yellow seeds.

When these first-generation plants were further bred among each other, many **AG**s and **AA**s arose in the second generation. All of these produced yellow seeds, as Mendel expected. However, notably missing from the first generation but present in the second were several **GG**s. In the absence of a dominating A, these **GG**s, a quarter or so of the total, produced green seeds. Indeed, the first-generation **AG**s secretly carried the memory of green in this character G, thus explaining Mendel's observations.

Coming back from the world of plants to *Dia* and her family—yellow and green are now replaced by two character variants, one which leads to vision loss (say, A), and the other which doesn't (say, G). Our quest is to identify this character among the billions of characters in the genome. This quest must keep the hide-and-seek behavior observed by Mendel in mind. In particular, it must be guided by the answer to the following question: does the variant A dominate the variant G? In other words, do **AG**s also suffer from vision loss, or does vision loss affect only **AA**s? We seek an answer to this question by studying the pattern of vision loss in *Dia*'s family more closely.

A Dominant Variant?

If indeed the problematic variant A dominates its benign counterpart G, its reflection should be apparent in *Dia*'s family. In particular, we should find that roughly half the children of an affected parent with **AG** are at risk of vision loss. This is so because roughly half these children will inherit the problematic A, while the other half will inherit the G and thus remain unaffected. Is this what we observe in *Dia*'s family?

Based on this hypothesis, we expect roughly four among *Dia* and her seven siblings to be affected by vision loss. The actual number, three of eight siblings, is only slightly lower. We should allow for some natural variation in these numbers; after all, eight fair coin tosses do not always yield exactly four heads. Therefore, a count of three among eight siblings appears consistent with the problematic variant A dominating its benign counterpart G.

How about the grandchildren *Dia*'s father had from his three affected children? For the same reason, we would expect half or more of these grandchildren to be affected as they grew into their 30s and 40s. Is that indeed true?

Dia's affected sister has three children, all well into their 50s, and none shows any signs of visual deterioration. *Dia*'s affected brother has three children, all in their 40s, only one of whom shows signs of being affected. *Dia* herself has two children, one in the early 40s and one in the late 30s; neither shows any signs of being affected.

For some of these younger grandchildren, it is probably too early to say. Nevertheless, the number of affected grandchildren appears to be far fewer than the expected 50%. This is not consistent with the dominant scenario—the character variant A which leads to vision loss appears unlikely to dominate its benign counterpart G. So, we turn to the other possibility—that G dominates A.

A Recessive Variant?

In the so-called *recessive* scenario, only individuals with **AA** suffer from vision loss; individuals with **AG** or **GG** do not. Such affected individuals must inherit an A from each parent; both parents must therefore be **AG** or **AA**. If **AG**, then they remain unaffected *carriers*, simply passing the problematic A along to the next generation.

We check if this scenario fits *Dia*'s family better. *Dia*'s parents were both unaffected, and therefore likely **AG**. Then, as in the second generation of Mendel's pea plants, *Dia* and each of her siblings would have a 25% chance of inheriting an A from both the parents, and hence suffering from vision loss. So, we expect a quarter of these siblings to be affected. The actual number, three out of eight siblings, is just slightly above this mark. This is broadly consistent with the recessive scenario.

How did the next generation after *Dia* fare? *Dia*'s affected brother *B*, clearly **AA**, has three children, one of whom is affected. This makes his spouse a carrier, hence **AG**. There is now a 50% chance that each child of *B* inherits an A from both parents. Accordingly, we expect roughly half of these children to be affected. The actual number, one out of three children, is again not far off the mark.

Dia's affected sister *S* has many children, none of whom is affected, even though they are all in their 50s. This could be simply because *S*'s spouse is not a carrier at all; this ensures that none of their children can be **AA**. *Dia*'s children themselves are in their late 30s and early 40s and yet unaffected—maybe too early to call, or maybe *Dia*'s spouse is also not a carrier. In either case, the recessive scenario appears to be the more promising avenue to explore.

Therefore, we embark on our quest for the character that causes *Dia*'s vision loss by seeking a character variant A such that affected individuals have **AA**—they have the same aberrant character in both chromosomes of a pair.

Where Do We Look?

Our genome has as many as six billion characters, of which just one or two may be responsible for *Dia*'s loss of vision. To keep our search for the needle in this haystack manageable, we focus only on *genes*—stretches of recipe-carrying text embedded within the genome.

The number of genes is not small, about 20,000. Even more overwhelming is the vast ocean of text in the genome outside of these genes. However, we start by focusing on genes alone, for the functions of this vast ocean of text are far less well understood.

Even as we do so, we have to remember that the recipe for a gene does not appear in one continuous stretch of text in the genome. Rather, it is split over many text stretches called *exons*, separated by intervening text stretches called *introns*. Only exons carry the recipe, introns don't. There are roughly 200,000 exons in all the genes put together. Together, these exons constitute just 1–2% of the entire genome—a small fraction, but nevertheless, as many as 50 million characters in all. We extract and sequence just these exons from the genomes of *Dia* and some other members of her family.

Of course, you might remember from the first story that genome sequencing is a convoluted process. It chops up *Dia*'s genome into pieces, called *reads*, and poses to us a jigsaw puzzle which we must assemble with some intense data crunching on a computer. To this end, we take each read and search for a match in the reference sequence—the pooled genomic sequence of a few healthy individuals. In this process, we keep an eye out for both perfect matches as well as matches with slight alterations, for the genomes of any two individuals do differ slightly (and only slightly). This so-called *read alignment* algorithm anchors each read to its rightful place in the reference sequence, as illustrated in this example.

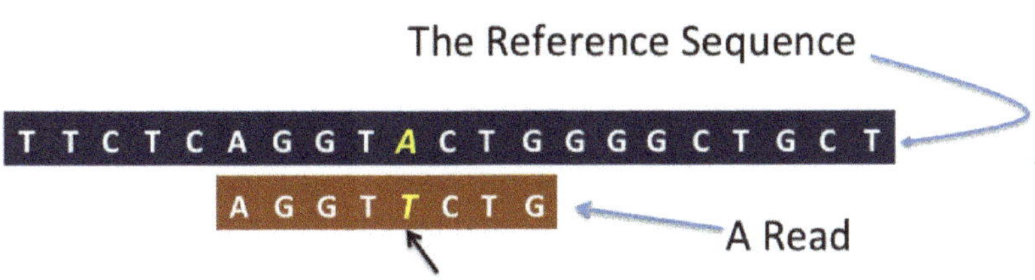

A read from Dia's genome anchored to the right place in the reference sequence. This is a match with one character altered.

Once each read from *Dia*'s genome is anchored to its right place in the reference sequence, differences between her genome and the reference sequence become apparent. For instance, in the picture above, *Dia*'s read shows a T while the reference sequence shows an A. This signals a difference between her genome sequence and the reference sequence. These differences are what we are interested in, for one of these differences might be the cause of *Dia*'s vision loss. We do have to be careful though, for what we see in the picture above might not be a true difference; it may simply be *measurement error.*

Most measurement methods carry small amounts of error; the sequencing process is no exception. It reads one in a few hundred characters wrongly. This is a low error rate, but far from negligible, since we are dealing with as many as five–six million differences between *Dia*'s genome and the reference sequence. Unless we correct for this error, tens of thousands of these differences will not be true differences, and we will be misled in our conclusions. So, we must carefully eliminate such errors. But how?

The trick lies in sequencing not just one copy of *Dia*'s genome but many copies simultaneously, as in this picture.

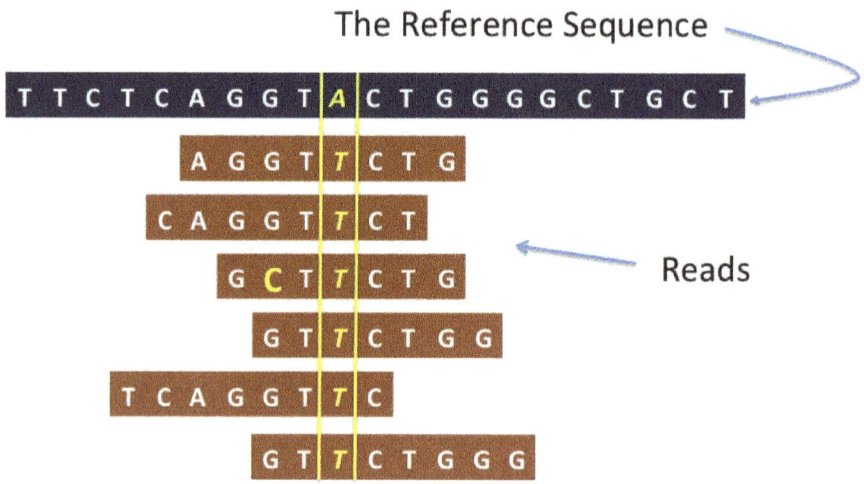

Multiple overlapping reads, all of which have a T while the reference sequence has an A. These reads together indicate a true difference between Dia's genome and the reference sequence. Note that one of the reads has a C where all others as well as the reference sequence have a G. That is an example of reading error.

Each of *Dia*'s cells carries the genome; we take several cells together, chop up the genomes in all these cells into reads, obtain the sequence of each read, and finally anchor each read to its rightful place in the reference sequence using the read alignment algorithm. We now have many different reads overlapping any given character in the reference sequence. If most of these reads indicate a T while the reference shows an A, as in the picture above, then *Dia*'s genome must truly have a T at this location. After all, errors are unlikely to occur in so many reads simultaneously.

However, as we proceed with this process, we also notice several instances where only half the reads or thereabouts indicate a T while the rest indicate an A, as below. How does this happen?

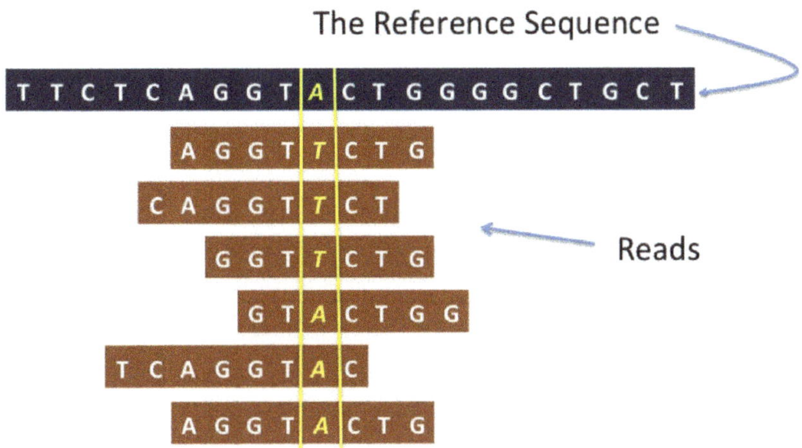

Roughly half the reads have a T while the reference sequence has an A. Why?

If we have a large enough number of overlapping reads, then reading error is not a plausible explanation, for reading error is unlikely to manifest in so many reads simultaneously. A moment of thought yields another, more likely, explanation.

Remember, most chromosomes come in pairs and the character sequences within a pair are similar but not identical. It must be that one of *Dia*'s two chromosome versions has a T, while the other has an A.

Proceeding further, we also encounter instances where very few reads indicate a T and most indicate an A. In these instances, both of *Dia*'s chromosome versions must have As, and the Ts are probably just measurement errors that can be ignored.

Thus, by analyzing reads piled up at every character in the reference sequence, we identify true differences between *Dia*'s genome and the reference sequence. For each such difference, or *variant* as it is called, we also identify whether it appears in one or both of her chromosome versions.

This done, we have at hand a list of *variants* for *Dia*. Presumably, the source of *Dia*'s vision loss lies in one of these variants. Our main problem though: there are tens of thousands of these variants. Which of these is the culprit?

A Glut of Variants

To sift our way through these tens of thousands of variants, we start by identifying genes of importance. Research spanning several decades has culminated in hundreds of thousands of scientific publications which tell us that not all 20,000 genes contribute to severe inheritable disease; only about 5000 of them do. This list could grow with time. Regardless, for the moment, these genes serve as a good starting point for our hunt.

We can restrict our hunting ground even further by considering only those genes that cause malfunction in the eye. Or even further, to genes which cause malfunction in the *retina*, the screen at the back of the eye on which our eye lens projects its image. A scan of published scientific literature for such genes yields a list of about 100 genes now, down from our starting list of 20,000 genes.

Our shortlist of genes spans a number of different retinal malfunctions. Some like *Leber Congenital Amaurosis* cause severe visual problems from early infancy, but are relatively rare, at one in 100,000 births. Others cause progressive loss of vision over several years and are more common. The most common among these is a disease called *Retinitis Pigmentosa*, with an incidence of one in 4,000. People with this disease have tunnel vision as shown in this picture.

Tunnel vision in Retinitis Pigmentosa.

Retinitis Pigmentosa is caused by loss of sensor cells in the retina—those cells which sense light. Remember, we encountered two types of sensor cells in the last story: *rods* and *cones*. Cones are more abundant at the center of the retina, in a region called the *macula*. Rods are more abundant outside the macula. With the onset of Retinitis Pigmentosa, rods are lost first, leading to tunnel vision of the form shown in the picture above. Cone loss follows later. More than 70 genes causing this disease are known.

In contrast, *Dia*'s vision was blurred more at the center, where cones are abundant, than at the periphery. This matches the typical pattern for diseases where cones are lost first, or both cones and rods are lost simultaneously. *Stargardt* disease is one such example, where degeneration of cells in the macula often starts in adolescence. *Best Vitelliform Macular Dystrophy* is yet another, with onset in childhood. Yet another but contrasting type of macular degeneration only sets in with age; it is called *Age-related Macular Degeneration* or *AMD* and typically appears in one's 60s.

In comparison, *Dia*'s vision loss started somewhere in between—in her 30s, making it hard for doctors to diagnose her condition. Hopefully, help is at hand—one of our tens of thousands of variants, now further restricted to about 100 genes, should give us the answer.

The Recipe Code

Our haystack is now smaller, but still sizeable, with 100 genes, each with potentially many variants. We prune it further by assessing how these variants disrupt gene recipes. A normal gene recipe may insist on a character G at a particular position, while a replacement with a character, say A, might change this recipe significantly. Which of the variants in our list disrupts a gene recipe significantly? The answer to this question depends on how gene recipes are written out in the first place. How, indeed?

These recipes are written out in genes using the so-called *genetic code*, discovered in the 1960s. To appreciate this code, pick a gene of your choice. Since its recipe lies in its exons, bring these exons together after excising out any intervening introns. Now divide the characters in these exons into triplets, or blocks of three characters each. Each triplet stands for one of twenty special molecules called *amino acids.*

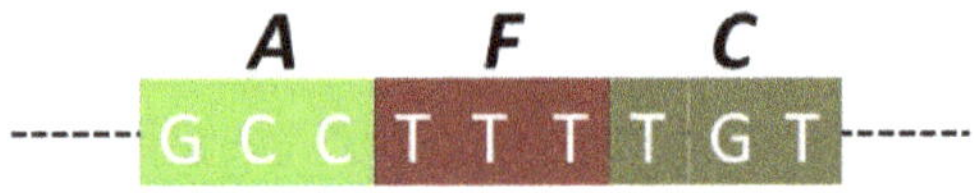

Triplets and their corresponding amino acid symbols.

These twenty amino acids do have proper names but often go by just alphabet symbols: *A*, *C*, *D*, *E*, *F*, *G*, *H*, *I*, *K*, *L*, *M*, *N*, *P*, *Q*, *R*, *S*, *T*, *V*, *W*, *Y*. Of course, they differ in various ways. Some are large, some are small, some are neutral, others are charged, some like being exposed to water, others hate it, and so on. Combinations of amino acids generated by different gene recipes then yield a wide variety of behavior that our cells use to conduct their daily business.

For the curious, the full genetic code depicting all possible triplets and their corresponding amino acids appears in this picture.

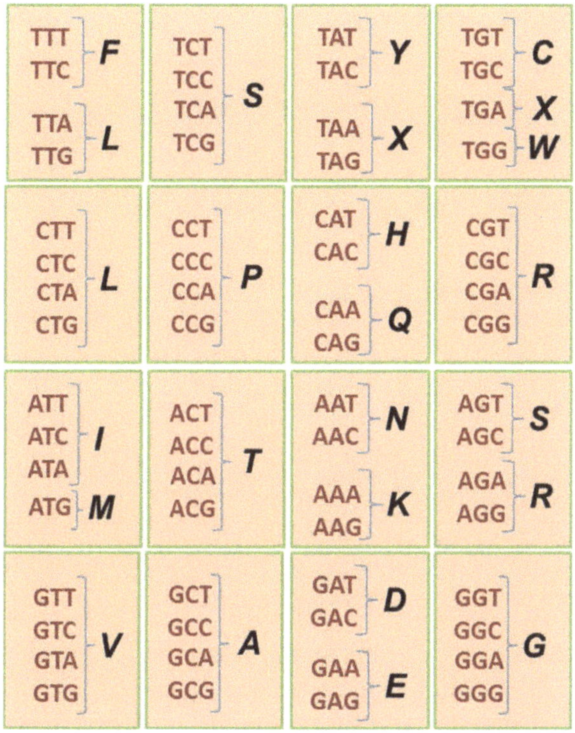

The genetic code. Amino acids are in bold italics. A few examples: triplet CTT specifies the amino acid Leucine (short form L), while triplets CCA and CCT both specify amino acid Proline (short form P).

In summary, what does a gene recipe yield? A recipe with 300 characters comprises 100 triplets. The recipe execution process simply takes each triplet in turn, creates the corresponding amino acid molecule, and strings these 100 amino acids together to produce the final product of this gene recipe—a single large molecule called a *protein*. This protein, which often carries the same name as the gene it came from, then participates in various chemical reactions needed for the healthy functioning of our cells. Different genes carry differing recipes and hence yield different proteins. As you might have guessed, the so-called protein in our diet provides us with adequate supplies of several of these amino acids, which are then strung together in varying combinations when gene recipes are executed.

Three triplets, *TGA*, *TAA* and *TAG*, deserve special mention at this point, for these are fundamentally different from the others. They don't code for any of the twenty amino acids listed above. Instead, they specify a *Stop* instruction. Carrying the short form *X*, this instruction serves to terminate the recipe execution process, thus declaring that the protein has now been fully created. Recipe execution stops when it encounters one of these three triplets.

Presumably, the source of *Dia*'s vision loss was a variant that created a major disruption in the recipe of its gene. What kind of disruption was this?

Synonymous, Missense, Nonsense

Imagine a variant in *Dia*'s genome—say, she has the character C instead of the usual T in the reference sequence, the genome sequence of a few supposedly healthy individuals. The triplet around this character now changes from, say, *CCT* in the reference sequence, to *CCC* in *Dia*. What happens then?

Fortunately, a glance at the genetic code indicates that both the triplets stand for the same amino acid *P*. So, nothing changes as far as the protein is concerned. Such so-called *Synonymous* variants cause no disruption to the recipe and are almost always silent and benign. Another glance at the genetic code should tell you that a modification to the last character of a triplet is often synonymous. Of course, biology is replete with exceptions—we will meet a little boy in a later story who has been put to great hardship on account of a synonymous variant. Regardless, *Dia*'s travails were unlikely to have been caused by such variants.

Not all variants are so benign though. For instance, imagine a variant which changes triplet *CCT* to *CTT*. This time, the amino acid changes—from *P* to *L*. *L* is a bigger amino acid than *P*, so its presence may cause the protein to behave very differently—it may stop participating in certain reactions that it normally would, or vice versa. The effect can then be dramatic on occasion, possibly even lethal, as we will see in a later story. *Dia*'s vision problems could well be on account of such a *Missense* variant.

There is yet another type of variant that engenders an even more dramatic effect. For instance, imagine a variant which changes triplet *AGA* to *TGA*. The amino acid *R* now switches to a *Stop* (short form *X*). Recipe execution, which is chugging along merrily, stringing one amino acid after another, suddenly halts in its tracks! The result is a partial protein, often quite incapable of performing its mandated duties—a dramatic outcome for a small A to T spelling error. Such so-called *Nonsense* variants could also underlie *Dia*'s vision problems.

There are a few other types of variants as well; we will encounter these in the stories that follow. For this story, we take our 100 genes that cause malfunction in the retina, and prune our list of variants to only missense or nonsense variants in these genes. Our haystack is much smaller now—yet it has several tens of variants. Hopefully, one or more of these holds the key to *Dia*'s puzzle. Which one(s)?

Common and Rare Variants

Missense variants are quite a mixed bag. Several are benign, but a few may have serious consequences. Elaborate experimentation is needed to separate the serious from the benign. In these experiments, proteins are created in semi-artificial systems with one amino acid replaced by another. The effects of this replacement are then observed. Doing this for the tens of candidates at hand is laborious and expensive. Instead, we take the easy way and lean on similar experiments that nature performs in the background all the time.

Indeed, nature is constantly experimenting with new genomic character combinations. Your genome is different from mine—so all of us are guinea pigs in this massive experiment. Now, if several healthy individuals have a particular missense variant in their genomes, then this variant is likely to be benign—otherwise, such a variant would have made its presence felt widely by causing vision loss on a larger scale. Such common variants can be identified by simply looking up genome databases comprising tens of thousands of healthy individuals. We prune such common variants away from our list. This leaves us with just a handful of rare variants in *Dia*'s genome to focus upon. Ironic as it might sound, information from genomes of healthy individuals goes a long way in helping an affected person like *Dia*.

But wait, did we not argue earlier that a recessive scenario is most likely—that the offending variant must be present in both chromosomes of a pair. Remember, we inherit one of these chromosomes from our father and the other from our mother. It is rare enough to have a rare variant in one of these two chromosomes, let alone in both. Unless, of course, the mother and father are themselves genetically related, or *consanguineous*. It is common practice in some communities in southern India for a woman to marry an uncle (her mother's brother). Children born of such parents are at increased risk for carrying the same rare variant in both chromosomes. We will return to this theme a little later in this story, for there are instances of consanguinity in *Dia*'s family as well.

Our hunt for the culprit is now nearing its end. We look in our shortlist of rare missense and nonsense variants for those which appear in both versions of *Dia*'s genome. And how many candidates do we find fitting this bill? Unfortunately, none!

The Culprits?

Were we wrong about the recessive scenario? We look again at our shortlist of rare missense and nonsense variants to see if anything interesting catches the eye. There absolutely isn't a variant in this shortlist that appears in both chromosomes of a pair. However, there are two variants which do catch the eye, simply because they both appear in the same gene.

This gene is located on chromosome 1 and goes by the name *ABCA4*. The first three letters of the English alphabet are coincidental; here, they actually stand for *ATP-Binding Cassette*—we will see why, shortly. *Dia*'s affected siblings, *B* and *S*, too have both these variants, as does *B*'s affected daughter *D*. This makes these variants even more interesting.

However, each variant appears in only one of the two versions of chromosome 1. Neither appears in both. This is puzzling. Could these variants still be the cause of *Dia*'s vision loss? Our quest for the answer to this question takes us through a tour of how we actually see.

How we See

The retina has special rod and cone sensor cells. These convert light to electrical signals, which are then carried to our brains by neurons. For *Dia*'s story, the scene of action revolves around this conversion of light to electrical signals in the rods and cones shown below.

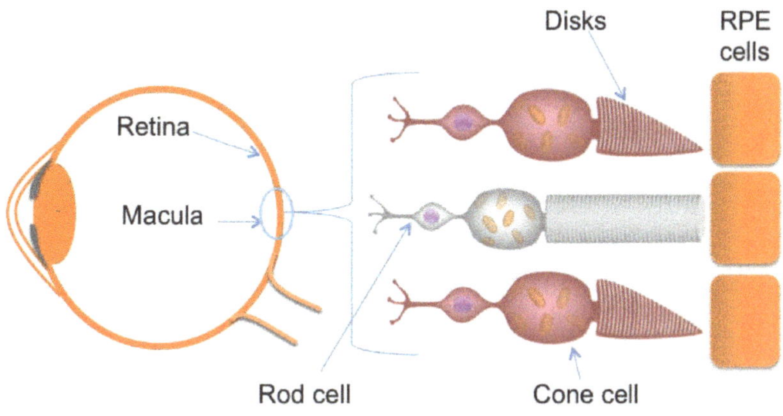

Rods and cones in the retina laid out on a supporting layer of RPE cells.

Note how these elongated rods and cones are placed on a supporting foundation of the so-called *RPE* cells. The scene of action, more specifically, is a stack of disks in these rods and cones abutting the RPE cells. These disks appear as striations in the picture above because they are shown in profile. They serve are nature's garbage bags, designed to package waste material for disposal. Just like garbage bags, used disks are shed and fresh new disks take their place. Discarded disks are eaten up by the RPE cells, thereby cleaning up unwanted debris. It is on and inside these disks that *Dia*'s story unwinds.

A single disk in close-up appears as in this next picture below—a hollow interior enclosed by a membrane.

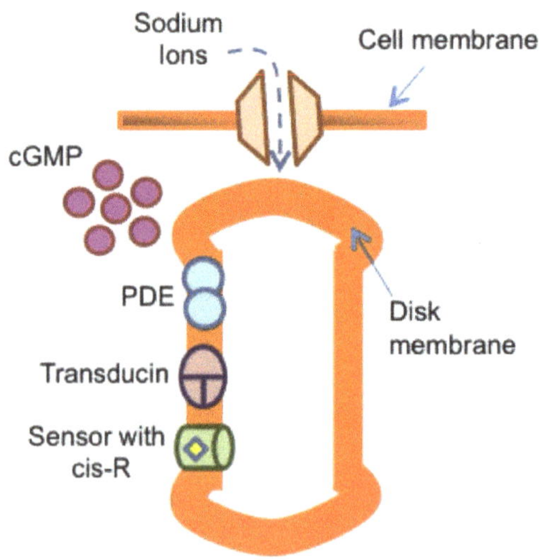

The cross-section of a single disk with the sensor molecule and other molecules on its membrane. Note the gate at the top is open, allowing the flow of electric current into the cell.

On this membrane are embedded various molecules that help us see. One of these is the light sensor molecule itself, made from recipes stored in the red, green and blue sensor genes we met in the previous story. Attached to this sensor is a little tag made from the Vitamin A in our diet, called *cis-R* (short for cis-Retinal)—hence the importance of Vitamin A for good vision.

Cis-R is the portal through which light is detected by light sensor molecules. It is hard to imagine light twisting an object out of shape, but that is exactly what light falling on this sensor does to cis-R. This twisted shape, called *trans-R*, can no longer stick to the sensor and comes off. The sensor molecule re-adjusts its own shape in response. What follows next is a cascade of several events.

The details of this cascade are numerous but not terribly important at the moment. It suffices to say that several molecules move around, react with others, or are destroyed in this cascade, eventually causing the gate that lets electric current into the cell to close. This is the critical step—electric current can no longer enter the cell through this gate. This signal of change in current is then transmitted to the brain by nerves.

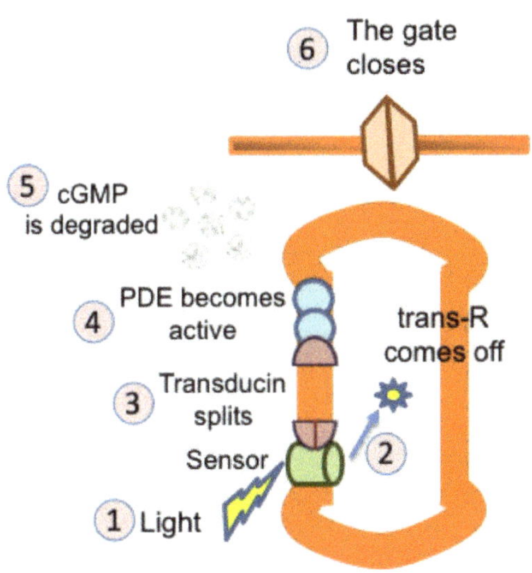

This cascade of events closes the open gate at the top after trans-R comes off the sensor molecule in response to light.

This is how light signals are detected, via cascades of interacting molecules. Our interest here, from *Dia*'s perspective, is not this process itself, but some collateral damage caused by cis-R and trans-R.

Collateral Damage

Light turns cis-R into trans-R. With light continuously bombarding the eye, more and more cis-R gets consumed. How does the eye continue to detect further impulses of light once all the cis-R is used up? A recycling mechanism converts trans-R back to cis-R for this very purpose.

After trans-R detaches from the sensor and initiates the cascade that closes the gate, it falls off, either outside the disk or inside. Outside the disk, molecules called RDH convert trans-R back to cis-R. Unfortunately, there are no RDH molecules inside the disk to mop up and recycle the trans-R that goes inside. So, any trans-R that falls off inside simply gets trapped, as shown here.

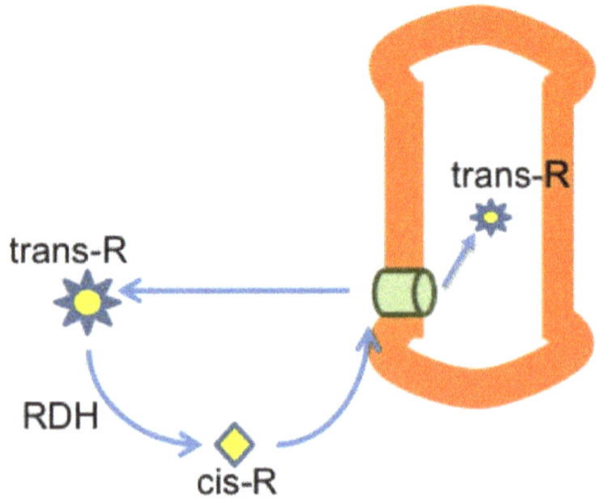

Trans-R outside the disk is recycled back to cis-R. The rest is trapped inside the disk.

The more light that falls on the retina, the more trans-R accumulates inside the disks. As these disks age and are eaten up by the surrounding RPE cells, trans-R and its derivatives accumulate in these cells. Some of these derivatives are degraded safely by the RPE cells. However, some others[7,8] stay undegraded, leading to toxic pile-ups.

A single RPE cell is estimated to gobble up three billion disks over 70 years.[9] This leads to a massive toxic pile-up which eventually kills these RPE cells. This debris is very clearly visible as a yellow fluorescent pigment when photographs of an affected retina are taken using special cameras.

As RPE cells die, rods and cones supported by these RPE cells also die. As rods and cones die, vision is lost. Cones seem to be particularly vulnerable to this toxicity, generating more toxic material and clearing less of it.[10] Therefore, vision is lost more at the center, where cones are concentrated, than at the periphery.

Of course, unlike *Dia*, most of us do not suffer vision loss. Something comes to our rescue. What?

ABCA4 to the Rescue

Something must be defusing all the toxic trans-R that accumulates within disks. Could it be a protein produced from the recipe in one of our 20,000 genes?

Indeed, it turns out that our savior is a protein produced by the very same gene that caught our attention when we sifted through the variants in *Dia*'s genome: *ABCA4*. This picture shows why.

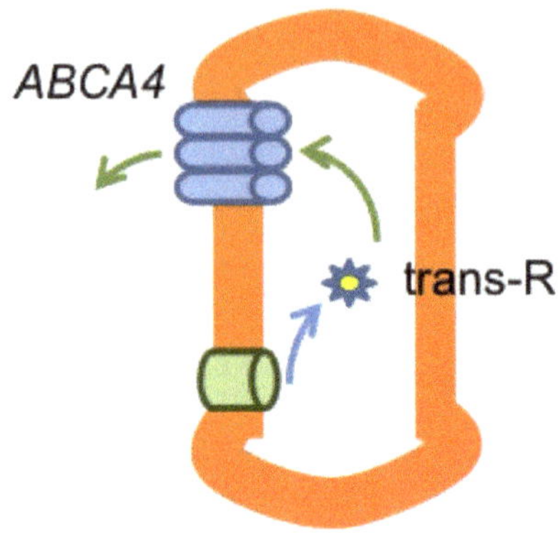

ABCA4 pumps out the trapped trans-R.

Like all other genes, the *ABCA4* gene provides the recipe for the creation of a specific protein molecule. Only cells in the retina execute this recipe, creating a chain of 2273 amino acids as a result. This chain folds into a specific three-dimensional shape to yield the *ABCA4* protein, which then moves to and parks itself on the disk membrane. Here, it performs a simple but very crucial function: it pumps the trapped trans-R out to be recycled. Toxic material no longer accumulates inside the disks, and our vision stays healthy as a consequence.

What, if anything, prevented *ABCA4* from performing its cleaning-up duties in *Dia* and her affected siblings? Could it be the two genomic variants we had identified earlier in *Dia*'s genome? We seek a better understanding of how *ABCA4* works to answer this question.

Powering *ABCA4*

We zoom into the site on the disk membrane where an *ABCA4* protein molecule is lodged.

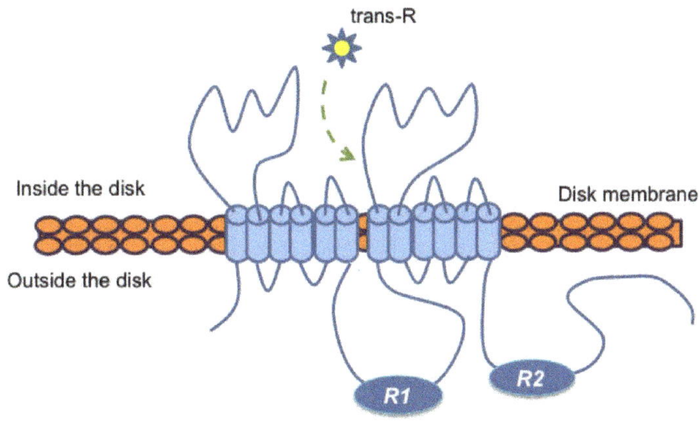

ABCA4 in the disk membrane, ready to pump out trans-R from the inside of the disk to the outside.

This chain of 2273 amino acids is shown as a long but stiff wire thread in the picture above as it weaves in and out of the disk. The parts of this wire that cut through the disk membrane are shown as thick cylinders. Two sections of the thread that appear outside the disk are particularly relevant to *Dia*'s story, and are highlighted with the labels, *R1* and *R2*.

Lodged in the disk membrane in this way, *ABCA4* is all set to pump trans-R out of the disk. This process starts with trans-R attaching itself to *ABCA4* inside the disk. Then, with a burst of energy, *ABCA4* drags it through the disk membrane and out of the disk. This energy is provided by a molecule called ATP—for which reason, the *ABC* in *ABCA4* stands for ATP-Binding Cassette.

ATP is nature's way of packaging energy from the food we eat into batteries so this energy can be consumed when and where needed. This ATP battery has three phosphate units; removal of one of these three units yields a burst of energy, while leaving only two units behind, and thus, the battery is discharged. Elsewhere, energy from food is used to replace the third unit, thus charging the battery once again for further use.

Coming back to *ABCA4*, ATP must attach to the portion labeled R2[11] to provide it with a burst of energy. This energy contorts *ABCA4*, forcing the R1 and R2 portions together, and dragging the attached trans-R through the disk membrane and out of the disk, as the contorted protein springs back to its usual shape.

So does *ABCA4* shepherd trans-R out of the disk, thus preventing the toxic build-up. *ABCA4* also cleans up any excessive cis-R in the disk membrane, for that too has similar toxic effects.[12] All of this is orchestrated by the contortions generated when the ATP battery attaches to R2. Did *Dia*'s genomic variants somehow make it harder for ATP to orchestrate this contortion?

Dia's *ABCA4* Recipe

Remember how *ABCA4* had caught our eye in the first place? Our shortlist of rare missense and nonsense variants in *Dia*'s genome had two variants which made her *ABCA4* recipe different from most others. The picture below shows the recipe change caused by the first of these two variants.

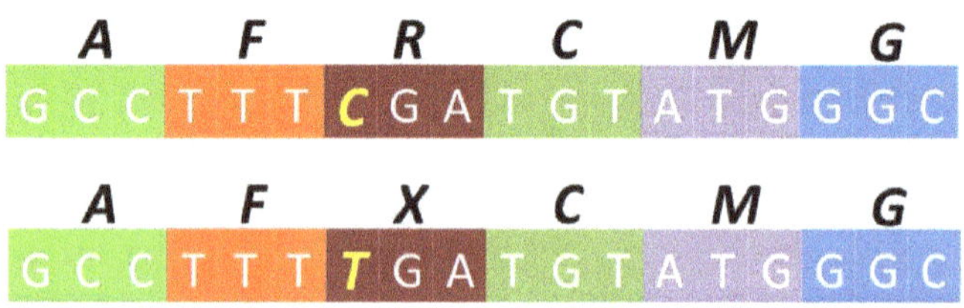

The recipe for most people is at the top, with triplets translated to amino acids. Dia's recipe is at the bottom. Amino acid R changes to X in Dia: a nonsense variant.

Dia's genome has a T where most others have a C. This seems innocuous enough, but the consequences are drastic: the triplet containing this character changes from an *R* to an *X*—the latter, if you remember, was short for a premature halt to the recipe. Typically, the *ABCA4* gene recipe yields 2273 amino acids before the recipe is halted. However, this variant brings the *ABCA4* recipe to a grinding halt at the 2149th amino acid itself.

Unfortunately, for *Dia*, the 2149th amino acid is located within the portion where ATP attaches to provide *ABCA4* the energy needed to drag trans-R across the disk membrane. With this portion truncated prematurely, it is unlikely that ATP can attach and fuel the transport of trans-R across the disk membrane. Even worse, such partial proteins are often identified by our cells as aberrant, and then sequestered in a corner and degraded. So, the protein created from *Dia*'s *ABCA4* recipe may not even be able to make its way to the playing field to play its role of pumping out trans-R.[13]

It appears likely that this genomic variant in *Dia*'s recipe, which we call R2149X, cripples *ABCA4*. R2149X has also been found in the genomes of other individuals with similar vision loss,[14] thus adding to the evidence and making it the likely cause of *Dia*'s vision loss. But there is a catch.

The *ABCA4* gene appears on chromosome 1. Remember that chromosomes come in pairs. So, a version of the *ABCA4* gene recipe is present in each member of the chromosome 1 pair. These two versions of the recipe are very similar, but not identical. Only one of these two versions carries the R2149X variant in *Dia*; the other does not. Indeed, this is what foxed us when we shortlisted variants in *Dia*'s genome—we didn't find any rare missense or nonsense variants that appeared in both versions of the recipe as expected in a recessive scenario. Therefore, wouldn't the other version of the *ABCA4* recipe yield a protein that holds the fort by pumping trans-R out of the disk?

Dia's Other *ABCA4* Recipe

Indeed, we have a conundrum on our hands. When we examined the pattern of loss of vision in *Dia*'s family, we concluded that the disease appears to follow a recessive scenario—both versions of the gene must be compromised by an aberrant character for vision loss to occur. However, we couldn't find any variant that was present in both gene versions in Dia. What are we missing?

We are missing the fact that the two versions of *ABCA4* could be compromised by two different variants! Indeed, our shortlist of rare missense and nonsense variants in Dia's genome comprised two distinct variants in *ABCA4*. The first variant in our shortlist, R2149X, compromises one version of *ABCA4*. We now need to verify that the second variant in our shortlist, also shown in the picture below, indeed compromises the other version of *ABCA4*.

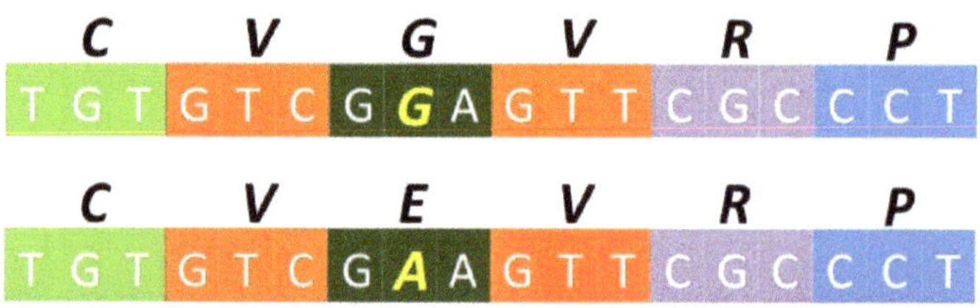

The recipe for most people is at the top, with triplets translated to amino acids. Dia's recipe is at the bottom. Amino acid G changes to E in Dia: a missense variant.

Of *ABCA4*'s 2273 amino acids, amino acid 1961 is usually a *G*. In Dia and her affected siblings, a genomic variant changes this recipe to *E*. This is a missense variant—one amino acid has been changed to another. Surprisingly, as many as five in a 1000 people are like Dia on this count: they too have an *E*. In India, the number is even higher, three in 200.[15] However, most of these people do not suffer from vision loss, because the variant is present in only one of the two gene versions; the other version is fully functional and holds the fort. In contrast, Dia's other version too was compromised, by R2149X.

Like R2149X, G1961E is also located within the portion of *ABCA4* where ATP attaches to provides energy. Its presence in two in 1000 individuals has prompted further research, which has shown that it reduces the ability of ATP to attach and power *ABCA4*.[16] However, unlike R2149X, it doesn't annul this ability altogether; it merely reduces it. In spite of this milder effect, the two genomic variants together compromise both versions of *ABCA4* to the extent that toxic trans-R accumulates over three decades.

So, this is how the simplest of spelling errors in *Dia*'s genome, one where a C is replaced by T, and another where a G is replaced by A, together gradually deprived *Dia*'s eyes of clear vision. What do these spelling errors mean for *Dia*'s extended family?

What This Means

Dia, her affected siblings *B* and *S*, as well as *B*'s affected daughter *D*, all have the G1961E and the R2149X variants—one variant in each of their two *ABCA4* gene versions. An ophthalmologist would accordingly classify their condition as Stargardt disease, named after Karl Stargardt, a German ophthalmologist who first characterized this disease in 1909.

Stargardt disease typically starts in the second decade of life and is therefore also called *juvenile macular dystrophy*. In *Dia*'s family, the relatively mild nature of G1961E pushes out the age of onset into the 30s and 40s instead. In contrast, had R2149X been present in both gene versions, the onset would very likely have been sooner.

There is good news for *Dia*'s children though. One is in the late 30s and the other is in the early 40s, roughly the age at which the disease is known to strike visibly. Fortunately, genome sequencing shows that both carry only the G1961E variant, that too only in one gene version. The other gene version is intact—so, neither is at risk of vision loss on this count.

B's two unaffected daughters are both in their 40s and have no symptoms thus far. One of these daughters has G1961E in both her gene versions. Is she also susceptible to vision loss? Remember, G1961E is a milder variant—it doesn't altogether stop ATP from powering *ABCA4*. Of course, it does cause vision loss when coupled with a starker variant in the other gene version. However, observations on patients with G1961E in both gene versions indicate a milder form of the disease, with widely varying ages of onset, ranging from 19 to 64.[17] So, this daughter of *B* may yet see some impact on vision, albeit mild.

Of course, R2149X in both gene versions would be far more serious in comparison. Fortunately, no member of *Dia*'s family has this combination. One of *Dia*'s nieces, *N*, has R2149X in one of her gene versions. She remains unaffected, for her other gene version is clear. She is a *carrier* though, and could have passed R2149X on to her children. Are they also at risk of developing Stargardt disease?

Each daughter of *N* has a 50% chance of inheriting R2149X from *N*. In addition, for vision loss, they must inherit a compromised gene version from their father as well. Usually, the father is unlikely to carry a compromised gene version, for both G1961E and R2149X are very rare—found in only three in 200 people in India. However, not so unlikely given a special circumstance in this family—*N* and her husband are genetically related first-cousins.

N and her husband form what is called a consanguineous couple. *Dia*'s extended family has many instances of G1961E and R2149X. By consanguinity, *N*'s husband has a far higher chance than three in 200 of carrying R2149X or G1961E—in fact, if you do the calculations, you get as high a chance as one in three, or about 33.3%. So, each of *N*'s daughters has a 16.6% chance of inheriting a problematic variant from their father. Coupled with the 50% chance of inheriting R2149X from their mother, this yields a net 8.3% chance for Stargardt disease. Not large, but not that small either.

N's daughters have not been tested for variants in the genome; neither has *N*'s husband. The dangling sword—an 8.3% chance of disease—could easily be removed by checking their *ABCA4* gene for variants. With more than 90% chance, this test can set that dangling sword to rest forever. However, there is a chance that the test might indicate otherwise, putting added psychological burden on *N*'s daughters. What can they do to delay the disease, if not prevent it, in that case?

Reducing exposure to light might reduce the rate of accumulation of trans-R inside the disk, thus postponing disease onset. Indeed, experiments done in mice kept in total darkness show reduced toxicity.[7] So, avoiding direct exposure to sunlight, using sunglasses, for instance, is potentially useful.

Conventional wisdom also prescribes the use of antioxidants like Vitamin A for good vision. But cis-R and trans-R are all derivatives of Vitamin A and increased Vitamin A consumption could lead to greater toxic accumulation in Stargardt patients. So, avoiding Vitamin A supplements is potentially useful as well. These measures could slow the disease down but are unlikely to stop it in its tracks.

Are there stronger measures that could help $\mathcal{N}$'s daughters, if they are indeed susceptible? Or restore vision loss in *Dia* and her affected siblings? Unfortunately, not yet. However, help might be around the corner, though it may be several years before it becomes available routinely.

Gene Therapy

If one is born with two defective versions of the *ABCA4* gene, can one compensate by injecting good versions of the gene into the eye? *Gene Therapy* intends to do exactly that.[18] There are many challenges though.

Injected *ABCA4* genes need to reach the sensor cells intact. This is not easy in general. Fortunately, the retina is easily accessible and genes can be directly deposited on specific cells in the retina via an injection. Reaching a more deeply buried organ like the lung would be much harder.

Next, sensor cells must be able to execute these injected recipes. The cleverness of gene therapy lies in its use of specially engineered viruses for this job. Some viruses have naturally evolved to attach themselves to certain types of human cells and let their genomes loose into these cells. Some go even further and integrate their genomes into the human genome itself. Human cells can no longer distinguish between human and viral genes, and execute both recipes. The *HIV* virus which causes *AIDS* is an example of such a virus. Of course, such viruses also cause undesirable ailments. So, safer versions of these viruses have been engineered with good *ABCA4* recipes packaged into their genomes for delivery into sensor cells.

Next, what if the virus incorporates its genome into the human genome at an inopportune location? For instance, in the middle of an important gene? There is danger then of disrupting that gene's recipe, with possibly serious consequences. So, special viruses that do not integrate their genomes into the human genome are used. These viral genomes are kept as separate entities called *episomes* inside human cells, still available for recipe execution but without disrupting the human genome.

Then there is the problem of dividing cells. Many cells in the human body are constantly dividing. When a cell divides, it breaks into two new cells. Episomes in the dividing cell then get split among the two daughter cells, reducing the number of episomes in any one cell. Over many rounds of division, few or no episomes will be left per cell. So, a single viral injection may not suffice to provide good *ABCA4* recipes to all the relevant cells; repeated viral injections may be needed. Fortunately, cells in the retina do not divide actively, so a single viral injection indeed suffices to provide a sustained supply of good *ABCA4*.

Finally, there is the issue of rejection and side effects. The human immune system often identifies and destroys foreign viruses, causing inflammation and scars in the process. This is one of the key stumbling blocks of gene therapy. The retina is however blessed in this regard; it can tolerate viral injection without a strong immune response.

So, diseases of the retina are particularly well-suited for gene therapy. The process of testing a therapy and readying it for routine medical use is quite slow and elaborate though. New therapies are first tested in animals. Trials in mice with severely compromised *ABCA4* genes have shown that a single injection of healthy *ABCA4* packaged in a specially designed virus (called *Lentivirus*) reduces accumulation of toxic material down to tolerable levels, even a year after the injection.[19] These promising results have justified launch of human trials.

The main human trial underway goes by the code name *StarGen*.[20] Such trials proceed in phases. Phase I checks whether the therapy is safe for use. Phase II checks if it is effective in stemming vision loss. Both phases test the therapy in a small number of patients. Phase III measures how well the therapy works on a much larger pool of patients. Phase I/II for StarGen started in 2011 and is expected to complete in 2017. If all goes well, StarGen may be ready for medical use a few years hence.

StarGen is probably too late for *Dia* and her affected siblings. Gene therapy can only help live cells circumvent malfunctioning *ABCA4*; it cannot resurrect dead cells. Unfortunately, *Dia* and her siblings have already suffered much cell death.

B's affected daughter though is still in her late 40s and could enroll in the StarGen trial, which is indeed recruiting patients. The children of *Dia*'s niece *N*, with their 8.3% chance of Stargardt disease ten years from now, would do well to track their vision carefully. Sequencing the *ABCA4* gene at the earliest sign of disease will confirm that the disease is due to *ABCA4* variants. Hopefully, StarGen would be well established and ready to help if and when this happens.

But what if StarGen's clinical trial fails to deliver? Are there other alternatives on the horizon?

Cell Therapy

Fortunately, a Phase I/II clinical trial is also ongoing for a completely different form of therapy.[21,22] It seeks to replace the dead RPE cells with fresh, healthy RPE cells, and is aptly called *Cell Therapy*.

Cell therapy requires a supply of fresh, healthy RPE cells. Where does this come from? Finding a donor and extracting RPE cells from the eyes of that donor is cumbersome if not infeasible. *Embryonic Stem Cells* serve as an alternative and plentiful source. These stem cells are obtained from early-stage embryos by combining sperm and egg cells from donors in a test tube. Left in the embryo, these cells will eventually differentiate into different cell types: some cells will mature into heart cells, others into brain cells etc. However, when these cells are taken out of the early-stage embryo, they continue to stay in their raw, undifferentiated state. Then, by clever programming, i.e., manipulating certain combinations of genes, these undifferentiated cells can be forced to differentiate into RPE cells.

These RPE cells are then injected into the retina. Ongoing trials will determine whether this therapy is safe and effective, but preliminary experiences on two patients appear encouraging.[21,22]

Reading Through the *X*

There are other alternatives on the horizon as well. Remember, one of *Dia*'s *ABCA4* variants was R2149X—a triplet specifying the amino acid *R* had become an *X* instead, truncating the *ABCA4* recipe prematurely. This may just turn out to be a minor stroke of luck, for drugs which allow the cell to read through the aberrant *X* triplet and continue executing the recipe are under active exploration.

There is an elegance to this approach—in theory, it is applicable to any gene and any disease caused by premature recipe truncation. However, proof is needed from actual human trials, which, unfortunately, are underway not for *ABCA4* but for some other genes.

The *DMD* gene causing *Duchenne Muscular Dystrophy* in young boys is one such gene. Boys with this disease progressively lose muscle strength and end up paralyzed at a very young age. The *CFTR* gene causing *Cystic Fibrosis* is another. In this disease, sticky mucous accumulates in the airways, causing breathing difficulties, and eventually leading to lung failure. Trials in both cases are ongoing and do seem to show positive results, at least in some individuals.[23]

Unfortunately, trials for Stargardt disease are not yet on the anvil. Hopefully, the promise of this approach will be explored at some point in the future.

Wrapping Up

Dia and her affected siblings have been living with Stargardt disease for the last 30–40 years. Yet, a definitive diagnosis by sequencing their genomes has become possible only very recently. We now understand the cause of their affliction. A lot of unwanted debris is created as light constantly bombards the eye. Nature has evolved mechanisms to clean up this debris. But nature's carefully laid plans are sometimes thrown into disarray by variants in the genome. And this reminds us of the inevitable—if unwanted debris is not cleaned up and disposed of properly, it will take its toll.

Administering good *ABCA4* recipes or healthy RPE cells into the retina will hopefully provide a way to control the disease soon. Though too late for *Dia* and her affected siblings, *B*'s affected daughter may be able to arrest further disease progression with these methods. Unfortunately, the rarity of Stargardt disease makes such efforts economically unviable or uninteresting for most pharmaceutical companies; non-profit foundations and governments have to step in to fund these trials and speed them up. After all, even at one in 10,000 people, there are an estimated 500,000 people with Stargardt disease in the world today.

Dia's children can breathe easy though, knowing that they are not at risk. In the meantime, *G1961E* and *R2149X* will continue to survive and propagate unnoticed in carrier individuals in this family. Their presence will be felt again when they cross paths—either with each other or with other problematic variants. Hopefully, by the time this happens, gene and cell therapies would have entered routine medical practice.

Story 3

The Rhythm Goes Awry

In this story, the scene of action moves to the heart. *Tara*, a young woman in her mid-20s, was frequently breathless. This was rather uncharacteristic for her age. Tests showed that her heart was unable to pump out blood with sufficient force—a life-threatening condition called *heart failure*. Her twin sister, *Cara*, too had the same issue.

Both twins also had abnormal heartbeat rhythms. A normal heartbeat allows the heart's chambers to fill with blood before they contract. In contrast, faster or chaotic heartbeat makes the heart contract even before its chambers are full. Not enough blood is pumped out then, causing sudden unconsciousness. Death follows in minutes unless emergency medical aid is provided.

Accordingly, an *ICD* device (an implantable cardiac defibrillator) was implanted in both the twins. This device attempts to correct irregular heart rhythms by providing compensatory shock pulses. The issue of heart failure remained though; the heart was just not pumping enough blood, so the situation remained life-threatening. Sadly, *Cara* passed away due to sudden cardiac arrest in her early 30s.

Sudden cardiac death was not new to us when we took on this case. One of our first hires at Strand was *Ro*, a very talented scientist in his mid-20s. He was frail but appeared healthy otherwise. Three years of productive and often intense work betrayed no signs of trouble lurking underneath. Then increased irregularity in his heart rhythms forced him to take a few weeks off. He never returned, succumbing to sudden cardiac arrest.

Ro, *Tara* and *Cara* all show(ed) warning signs of irregular heart rhythms. Such signs are not always present; sometimes, this so-called arrhythmia can arise suddenly and dramatically. An example is the case of footballer Miklós Fehér, a Hungarian footballer playing for the Portuguese club Benfica. On 25 January 2004, 24-year-old Fehér came on as substitute in a match against the club Vitória de Guimarães, and assisted with scoring a goal. Toward the end of the game, in injury time, he collapsed to the ground. Emergency medical attention did not help. The cause of death was soon confirmed to be sudden onset of arrhythmia.

Apparently, *Tara*'s father had also passed away early, in his 40s. The exact cause of his death was not known; his family members remember it simply as a "heart attack". *Tara* also has another sister *S* and a brother *B*; both have no symptoms whatsoever. To be doubly sure, they were tested for heart function. And these tests revealed the unexpected. *B*'s heart appeared to be completely normal. In contrast, *S*'s heart showed the same dangerous heartbeat rhythm as *Tara*'s, even though she had not felt breathless nor shown any other symptoms. An ICD was implanted in her as a precaution.

What caused abnormal heart rhythms and heart failure at such a young age in this family? Could *B*'s currently normal heart rhythms turn abnormal over time? In addition, were the children of *S* and *B* at risk?

Where Do We Look?

Sisters *Tara*, *Cara*, *S* and their father, all very likely share a genomic character which causes these abnormal heart rhythms. Our quest is to identify this offending character by sequencing *Tara*'s and *S*'s genomes.

As in our previous stories, we focus only on genes—those 20,000 or so recipes embedded within the genome. Of course, these recipes are written staccato in the exons while skipping over intervening introns. Accordingly, we extract and sequence only these exons, about 50 millions characters in all. This gives us a large number of reads, or tiny shreds of *Tara*'s genome. We then put this jigsaw puzzle back together using the reference sequence, the genome sequence of a few supposedly healthy individuals. That done, we obtain a list of variant characters where recipes in *Tara* differ from those in the reference sequence.

Presumably, one or two of these variants hold the key to the mystery in this family. As before, our main problem is that there are tens of thousands of these variants. Which of these is the culprit? How do we identify this needle in our vast haystack?

We start by focusing on genes that play important roles in the heart. There is much scientific literature on this theme, stemming from research over the last few decades. A computer-assisted scan of this literature gives us a shortlist. Though limited by the state of current knowledge, this shortlist serves as a good starting point for our hunt.

There are 200 or so genes in this shortlist. What do these genes do? How do they keep the heart beating every second, lifelong? Which of these genes is the cause of *Tara*'s condition? Our quest for answers to these questions first takes us through a tour of the heart.

The Heart

The heart's job, in essence, seems particularly simple—to keep the blood in circulation at all times. In its simplest form, for instance in fish, blood flows in a single circulatory loop—one push from the heart propels it with sufficient force to make a round-trip through the body and all the way back to the heart. Along the way, it passes through the gills and gains oxygen which it delivers on its round-trip back.

For larger organisms, like us humans, this single-loop design poses a challenge. Blood vessels in the gills and lungs necessarily have thin walls to allow oxygen and carbon dioxide to easily diffuse through. The heart must therefore be sensitive and not pump blood with too much pressure, lest it damage these fragile walls. At the same time, it must pump with sufficient pressure to enable a round-trip back through every nook and corner of the body. This delicate balancing act doesn't quite work out for larger organisms. Indeed, such organisms, humans inclusive, have evolved two circulatory loops instead of one, as illustrated by the two colors in this somewhat busy picture.

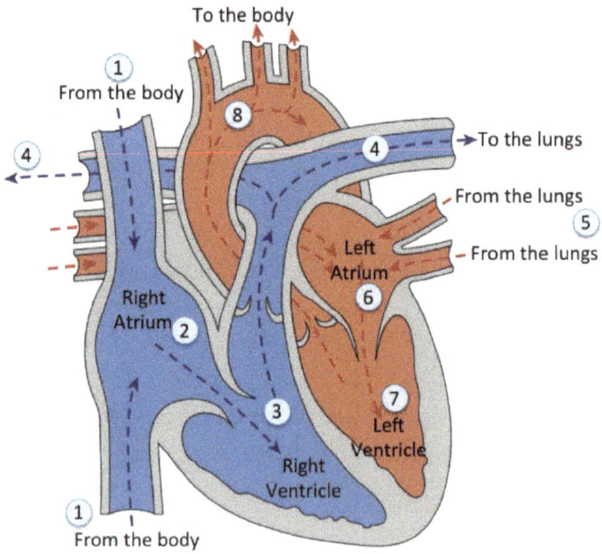

The heart with its four-chambered structure and two distinct blood flow loops. You can take a moment to stare at the various arrows shown above, or, alternatively, simply ignore these and read on.

In this two-loop design, one loop pushes blood to the lungs for oxygenation and back, while the other propels blood to the rest of the body and back. This separation allows the heart to be gentler on the first loop, but much more forceful on the second. Accordingly, our hearts have two halves. The right half, which is gentler, handles the first loop, while the more muscular left half handles the second loop.

Then it turns out that each half of the heart comprises not one but two chambers—an *atrium* and a *ventricle*, making four chambers in all. When ventricles contract, blood is pumped out of the heart. Of course, what is pumped out must come back at the other end; this incoming blood is welcomed by the atria, which are relaxed at this point. Once full, the atria contract, pumping their contents into the ventricles, whose turn it is now to relax. This two-beat rhythm, the atria and the ventricles contracting alternately, is what keeps our blood in circulation. Valves ensure that these contractions do not send blood rushing the wrong way.

Interestingly, in between the two-chambered, single-loop design in fish, and the four-chambered, two-loop design in humans, lies an intermediate—a three-chambered, two-loop design, found in frogs and snakes. This design has only a single ventricle where oxygenated and deoxygenated blood is mixed up. In contrast, humans and other mammals, and surprisingly even birds, maintain a clean separation of oxygenated and deoxygenated blood using their four-chambered design, and this makes for more efficient oxygen delivery.

Regardless of what the design might be, the heart in all organisms has one key characteristic—it must beat day in and day out, steadily and tirelessly, for years or decades at end, without a pause, for a pause of even a few minutes can prove fatal. What keeps these contractions going 24/7, and which of these elements could have been compromised in *Tara*?

The Heartbeat

Let us start with the question: what sets the rhythm for these steady, tireless contractions of the atria and the ventricles? A surprising observation gives us a hint: the heart continues to beat for some time even when it is cut off from the body! Something within must be setting the rhythm. Indeed, it turns out that a bundle of cells in the top-right corner of the right atrium behaves like a clock, putting out a steady beat, to which muscle cells in the heart respond by contracting.

These clock cells have gates on their boundaries that let electrical current pass through. Like a pendulum swinging back and forth, these gates open and close in a certain systematic sequence. As gates open and close, ions rush in and out, causing the voltage difference between the inside and outside of the cell to oscillate. This oscillating electrical signal is the genesis of the heartbeat.

At least 12 genes, called *channel* genes, carry recipes for the creation of these voltage-sensitive gates. Each gene specializes in one type of electric current—comprising either sodium, potassium or calcium ions. The picture below shows how these gates open and close periodically at various stages of the heartbeat.

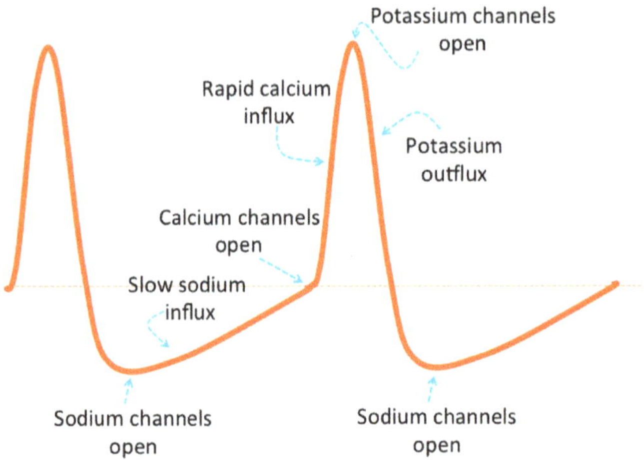

The electrical signal generated as ions rush in and out.

Each gate must open or close precisely when the voltage reaches a certain level. If it plays truant and doesn't open or close at the required moment, the heartbeat can go awry. Did variants in *Tara*'s genome modify one of these channel genes, so the corresponding gate did not open or close with the required clockwork precision?

Before we seek an answer to this question, another question poses itself. How is this heartbeat generated by the clock cells transferred to muscle cells all over the heart, so that they may contract in synchrony and pump out blood?

Of course, there are no wires in the heart. Instead, cells in the heart have special conduits between neighbors, which allow charged ions from one cell to move freely to the neighboring cells. When the voltage inside one cell increases, ions flow from that cell to its neighbors, thus raising the voltage inside the neighboring cells as well. The electrical signal is thus able to spread from cell to cell, as shown below.

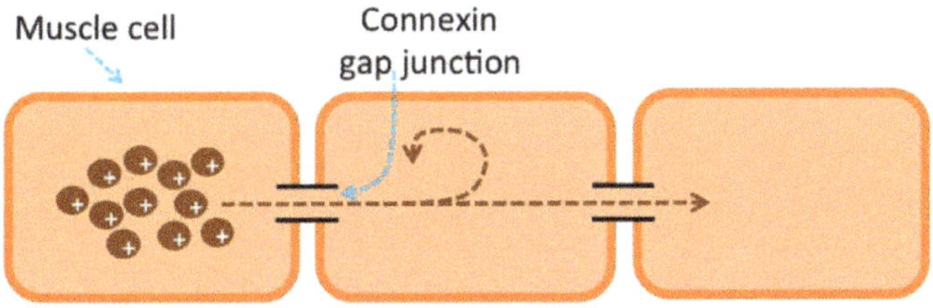

Conduits between cells transmit the heartbeat.

These conduits between cells are made from genes called *connexins*. Not all connexins behave the same way though. Some cells have conduits made from faster connexins while others have conduits made from slower connexins.[24] The heart needs both because not all parts of the heart contract simultaneously in response to the heartbeat. In fact, if you recall from a few moments ago, there are two separate waves of contraction—atria contract first and ventricles follow next. To orchestrate this sequence, fast connexins must rapidly disseminate the electrical signal from the clock throughout the atria, while slower connexins delay this signal from reaching the ventricles. Once this delayed signal reaches the ventricles, which happens after the atria have contracted, fast connexins must once again rapidly disseminate this signal throughout the ventricles.

As you can imagine, this careful orchestration could go for a toss if genomic variants caused some of these connexins to become faster or slower. Was this the case for *Tara*?

A brief digression, before we get back to *Tara*. Roughly 330,000 people die of sudden cardiac arrest every year in the United States. Mysteriously, 5% of these patients have absolutely no other signs or symptoms—no clogged arteries, no heart malformation, and no signs of age-related decay. The cause of death in these cases remains unexplained even after an autopsy and a comprehensive forensic investigation. Indeed, the needle in such situations might just point to faulty channel or connexin genes, as the following study indicates.

The office of the Chief Medical Examiner, New York City, studied 274 such deaths,[25] split almost equally between infants and people in the 19–58 age-group, all young enough that age-related problems of the heart were an unlikely cause. They checked each individual for genomic variants in just six of the several channel genes and found such variants in 16% of the cases. About half of these cases involved a single sodium channel gene, *SCN5A*. What is remarkable is that genome sequencing revealed the underlying cause of death, which an autopsy and forensic examination could not.

Coming back to *Tara*, could genomic variants in her channel genes or her connexin genes be the cause of her condition? Some thought suggests this is unlikely, because a scan of her heart showed added abnormalities—her heart muscle walls appeared distinctively weak. Problems in channel genes or connexin genes modify heart rhythms but usually do not weaken the heart muscle. Genes which do so may be a more likely cause, so let us look at those genes next.

The Muscle

How does the heart convert rhythmic heartbeat to powerful contraction? A clever contraption that listens to the heartbeat and contracts in response does the honors. Muscle cells in the heart are packed with these contraptions, called *sarcomeres*, as shown in the picture below.

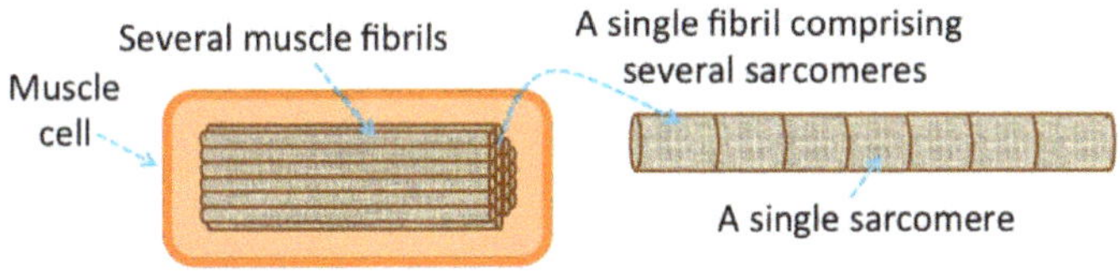

Bundles of sarcomeres in a muscle cell.

A sarcomere is a clever and complex piece of engineering. It has several components, of which two rope-like filaments, marked *actin* and *myosin* in the picture below, are key.

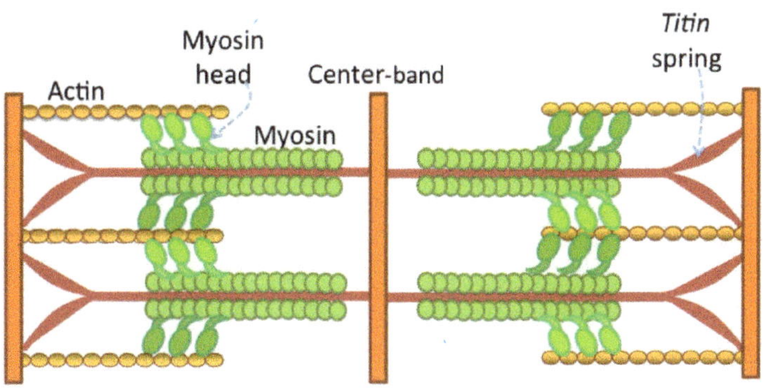

A relaxed sarcomere.

When the heartbeat signal arrives, myosin filaments pull on the actin filaments using their so-called *heads*, thus causing the ends of the sarcomere to be drawn toward the center, leading to the contracted form shown below.

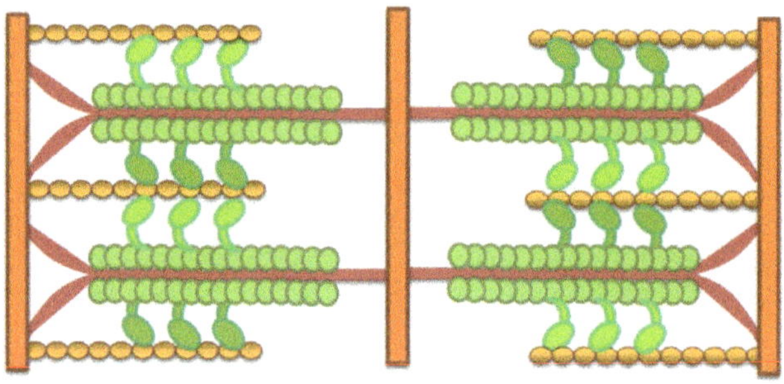

A compressed sarcomere.

Many different genes provide components of the sarcomere. Some contribute to the myosin and actin filaments. Others contribute proteins which prevent myosin heads from pulling on actin all the time. These proteins fall off when the heartbeat arrives in the form of calcium ions, paving the way for contraction.

Did you also notice from the picture above that the myosin and actin filaments themselves are not elastic; they do not compress in length when the sarcomere contracts. What compresses is *titin*, a spring that holds the sarcomere together. Its recipe resides in the longest gene in our genome, the *TTN* gene, with a whopping 34,350 amino acids. Remember, *ABCA4* from the story on *Dia*'s vision loss, had a mere 2273 amino acids in comparison.

Were variants in any of these sarcomere genes the cause of *Tara*'s condition? Indeed, faults in several sarcomere genes are known to result in abnormal heart muscle, which in turn makes heart rhythms go awry. *Hypertrophic Cardiomyopathy* (*HCM*, for short) is the most common of these abnormalities—the walls of the left ventricle become abnormally thick and stiff, as in the picture below.

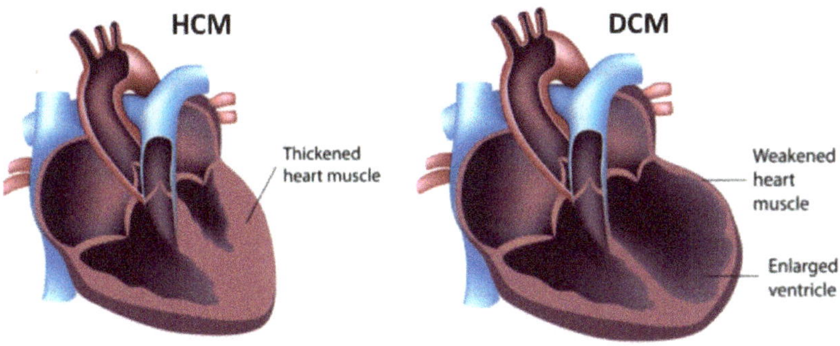

A hypertrophic heart and a dilated heart.

This thickening reduces the volume of blood that the left ventricle can carry and pump. Physical exertion can then bring about chaotic heart rhythms, which can be lethal on occasion, as seen in athletes like Fehér. HCM is found in as many as one in 500 young adults,[26] making it the most common cause of heart-related sudden death in people under 30.

Another condition called *Dilated Cardiomyopathy* or *DCM* is less common, at one in 2500 individuals. It is the opposite of HCM in a way—the left ventricle wall is thinner and weaker and the heart appears dilated, as shown in the picture above. Like HCM, it too is caused by variants in genes of the sarcomere.

Tara's heart presented a picture that resembled DCM more than HCM. So genes that contribute to the sarcomere were definitely a possible cause of her condition. The resemblance to DCM was not perfect though; there were some differences—for instance, there were fissures in the muscle wall of her left ventricle. Which other genes could contribute to something like this?

The Glue

You might now ask how rhythmic contraction of these individual sarcomeres translates to rhythmic contraction of the entire heart. After all, the sarcomeres could well pulsate in place without carrying the entire cell along, or different cells could just throb in place rather than all of them contracting together in unison. There must be some glue that binds the sarcomeres to the cells and the cells to each other, thus combining their individual contractions into a powerful combined contraction of the atria or the ventricles. Indeed, there is, as shown in the picture below.

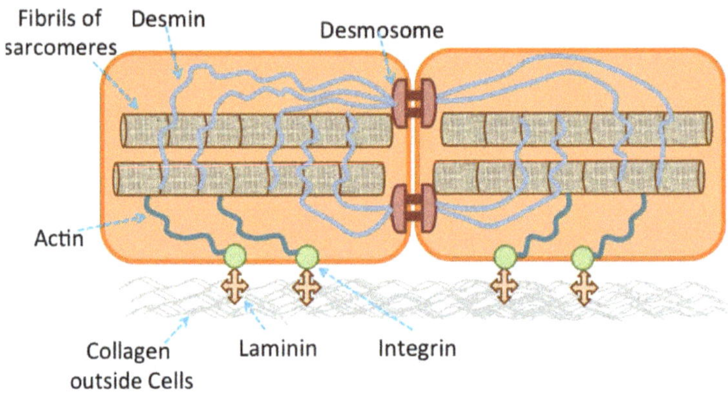

The glue that connects sarcomeres to cells, and cells to each other.

A muscle cell is clamped to its neighboring cells by rivets called *desmosomes*, made of proteins from five genes. Sarcomeres within a muscle cell are connected to these rivets and to various parts of the cell by a network of filaments (remember, ropes). The thicker filaments are made with *desmin* proteins, while the thinner ones are made with *actin* proteins. Yet other proteins anchor these filaments to a cushioning material called *collagen* that fills the space between the various muscle cells. All this glue forces the entire collection to contract together as one unified whole.

Just as a building subject to chronic high vibration levels develops cracks unless the cement used is strong, so does the heart suffer structural damage with years of relentless beating, unless the glue is strong. Variants in genes sometimes compromise this glue, paving the way for accelerated wear and tear of the heart.

A classic example are variants in genes that contribute to the rivets (or desmosomes). These cause a disease called *Arrhythmogenic Right Ventricular Dysplasia (ARVD)*, where heart muscle, typically starting with the right ventricle, gets progressively damaged and replaced by fat and fibrous tissue. This fat and fibrous tissue interferes with the transmission of the heartbeat, causing arrhythmias, and possibly death. More males seem to have ARVD than females.

Could genes involved in these glueing mechanisms be the cause of *Tara*'s condition? *Tara* and her affected siblings were all females. But that only makes them slightly less susceptible to ARVD than males. More importantly, structural damage appeared prominent in the left ventricle in *Tara*'s case, not quite matching the right ventricular focus of classic ARVD.

The picture at this point appears to be far from clear. The source of *Tara*'s problem could lie in the heartbeat (unlikely), in the muscle (possibly), or in the glue (possibly). This diagnosis seems difficult to make without diving into her genome. That is indeed our next step. But before we launch our hunt, we need to know what we are looking for—for a problematic variant that dominates its benign counterpart, or one that is dominated by its benign counterpart?

Dominant or Recessive?

Unlike what we saw with *ABCA4* in the story on *Dia*'s vision loss, many variants in genes involved in the heartbeat, the muscle, and the glue, behave in a dominant fashion—a problematic variant in just one of the two gene versions suffices to cause disease. This is odd, isn't it? When one version of *ABCA4* has a problematic variant and is unable to clear up toxic material, the other good version takes over and does the job. Why does the same not happen here?

Possibly, one good gene version doesn't generate enough good protein by itself to get the job done. An analogy might help here. If a building needs 100 pillars for strength, and the architect puts in only a 100 pillars, then the failure of half of these pillars definitely leaves the building compromised. The good pillars alone cannot rescue the situation.

Alternatively, the problematic gene version interferes with the functioning of the good version, spoiling it in the process. As an analogy, imagine 100 identical strings vibrating in unison and producing a certain pleasant note; if half of these strings go bad, then the result will be cacophonous, even though 50 of them are good. A mix of good and bad protein resulting from variants in one of the two gene versions could thus upset the various forms of careful synchrony in the heart we saw earlier. Here is one of the cases from our practice that illustrates this.

A man *Nero* in his 20s was diagnosed with DCM. His left ventricle was enlarged. Its walls appeared weak and thin. His heart could pump out only 20–25% of the blood in the left ventricle with every heartbeat. He was advised an ICD implant, much like *Tara*. Sequencing his genome indicated that he carried a variant in one of his *TTN* gene versions. Remember, this is the gene with the longest recipe in our genome; it contributes the spring that holds the sarcomere together. *Nero*'s variant resulted in a shorter spring with only 32,946 amino acids instead of the usual 34,350. This truncation was in only one gene version, the other was untouched. Yet this other version could not hold the fort for *Nero*. Why?

Truncated proteins are often identified and destroyed by cells. Not always though; they do slip through on occasion. Indeed, there is some evidence to suggest that *Nero*'s sarcomeres contained a mix of truncated and full-length springs,[27,28] compromising the sarcomere and leading to DCM. *Nero* was unlucky on this count by just a whisker; had the truncation been 500 or so amino acids further down the recipe, it may not have been severe enough to cause the disease.[27] Truncations would then be needed in both gene versions for the sarcomere to be compromised.

Nero's case is an example of how a problematic variant in just one of the two gene versions can cause disease. In this dominant scenario, a child has a 50% chance of inheriting the lone problematic version from their affected parent. So, one expects roughly half of *Nero*'s children to be affected. Fortunately, a rather heartwarming act of fate ensured that neither of his two young children, an infant and a toddler, had inherited the problematic variant. *Tara* and her siblings were less fortunate; three of four siblings, including *Tara*, seem to have inherited the problematic gene version from their father. This only reaffirms the dominant scenario. The stage is now all set for us to return to *Tara*'s variants and launch our hunt for the culprit gene keeping this dominant scenario in mind.

Sifting Through Genes

Our hunt returns to our shortlist of 200 or so genes, altered recipes of which are known to impact the heart. We keep in mind that the gene of interest for *Tara* is likely to be one that contributes to the sarcomere or to the glue. Accordingly, we look for recipe changes in these genes brought about by missense and nonsense variants—those that substitute one amino acid for another, or those that truncate the recipe prematurely.

We further focus on variants that are present both in *Tara* and in her lone surviving affected sister. Of these, we consider only rare variants, which are not commonly found in many people; after all, *Tara*'s condition is certainly not a common one. Continuing in this vein, and keeping the dominant scenario in mind, we look for variants that are present in even one of the two gene versions. With this, our shortlist reduces dramatically to the low single digits.

We wade through this shortlist. There is the *OBSCN* gene, which contributes components to the sarcomere. There are two missense variants in this gene. Both are very rare. We look to see if any other patient whom we have tested earlier has any of these variants. One other patient indeed has one of these variants. But this patient has an eye disease with no cardiac complications. The second variant has not been seen in any of our patients previously. It does seem to be in an unimportant section of the recipe though, so we eliminate this gene from consideration.

We move on to the *TTN* gene, the longest gene in our genome and the likely cause of DCM in *Nero*, whose case we encountered earlier. There are a large number of missense variants in this gene. This is not surprising; every person whose genome we have sequenced has many such variants. Typically, the longer the gene, the more the number of variants. And this gene is the longest of all. Missense variants in this gene are usually innocuous, but we cannot be sure. There are no variants that truncate the recipe prematurely though, of the type we saw in *Nero*. So, we keep this gene also aside for now.

Then there is an intriguing missense variant in the *LAMA4* gene. This gene helps anchor the filament network inside the cell to the cushioning material outside (remember laminin in our glue picture earlier). However, the variant appears to have never been encountered earlier, which again leaves us playing a guessing game about its effect. Only two other variants in this gene causing heart disease have ever been published in scientific literature, both in individuals with DCM.[29] So, is the glue weakened by our missense variant? We just don't have an answer.

This is exasperating, and much harder than we expected—assessing variants about which little is known. We are hunting for a needle in a haystack, that too in the dark. For the moment, we keep the *TTN* and *LAMA4* variants aside, making a mental note to come back and dig deeper if we cannot find a more promising candidate. We have only one candidate left, which we dig into next, with bated breath.

Missing Characters

The variant in our last gene candidate is unlike any we have seen so far. It is not as if a character in the reference sequence has been replaced by another in *Tara*'s genome. Rather, a couple of characters present in the reference sequence are completely missing from *Tara*'s genome. How do we know this? From the picture below.

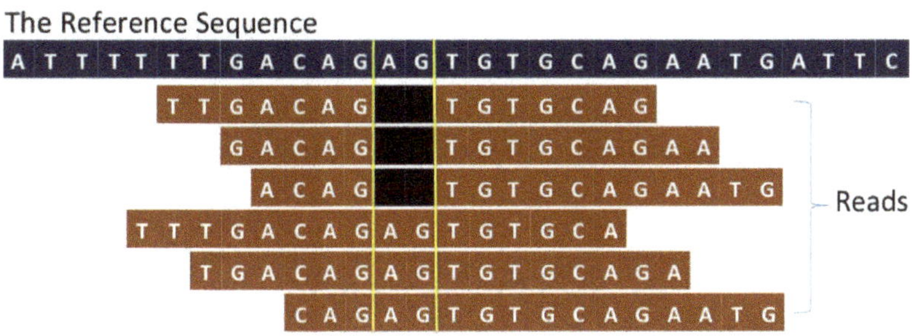

Two missing characters in Tara's genome.

This picture shows *Tara*'s reads—those tiny shreds of genomic text we obtain when we sequence her genome. From our first story on color blindness, you will recall that we assemble a jigsaw puzzle posed by these reads. In this process, each read gets placed at its rightful position in the reference sequence. Then, the missing characters become more than apparent. A glance at the picture above tells us that the characters AG are missing from half the reads. So, one of *Tara*'s two genome versions is missing these characters. The other appears to be normal.

Some care in solving the jigsaw posed by *Tara*'s reads is needed to get this far though. We must search for these reads in the reference sequence, keeping in mind that the reference sequence differs from *Tara*'s genome, even if only slightly—one in a 1000 characters on average. So, we must keep a keen eye out for close matches.

In *Dia*'s case of vision loss in the previous story, we got away by seeking close matches where a character or two is different between the read and the reference sequence. Here, we have to go further and allow for the read to have some extra characters and some that might entirely be missing. This is easier said than done though.

For instance, suppose we have a read, say, AGGTCCTG, from *Tara*'s genome. As we scan the reference sequence for matches, we might find, say, AGG<u>C</u>CCTG. Here, a T appears in the read instead of the underlined C in the reference sequence. Continuing further, we might find AGGT<u>T</u>CCTG—an extra underlined T appears in the reference sequence. Or even AGG_CCTG—the T is missing completely in the reference sequence, as indicated by an empty underlined slot. We must identify all these possibilities and locate the one amongst these that offers the closest match to the read.

This is tricky business and hard work—knowing which characters to modify, add or remove as one scans the reference sequence for each read. Of course, we use computer algorithms to do this. How does the computer know which characters it must modify, add or remove? Well, it doesn't. It just has to try all possibilities. Would that not take forever, that too with tens of millions of reads to process? Not really—clever algorithms enable a computer to breeze through this. We will see glimpses of this algorithm in a later story.

Coming back to *Tara*—with all this careful work, we determine that our last gene candidate has two characters missing in her genome when compared to most others. Could these two characters provide a clue to our mystery?

Recipe Out Of Frame

Remember gene recipes from the story on *Dia*'s vision loss? We saw how two rather innocuous-looking character substitutions modified gene recipes in unexpected ways, causing *Dia*'s vision loss. What recipe change do our two missing characters bring about in *Tara*?

We quickly take the gene with the two missing characters, string together its exons after excising out intervening introns, and divide the characters in these exons into triplets. Inclusive of the two missing characters, there are 2871 such triplets. Since each of these triplets specifies an amino acid, the whole recipe specifies a protein with 2871 amino acids. The picture below shows how the two missing characters affect these amino acids.

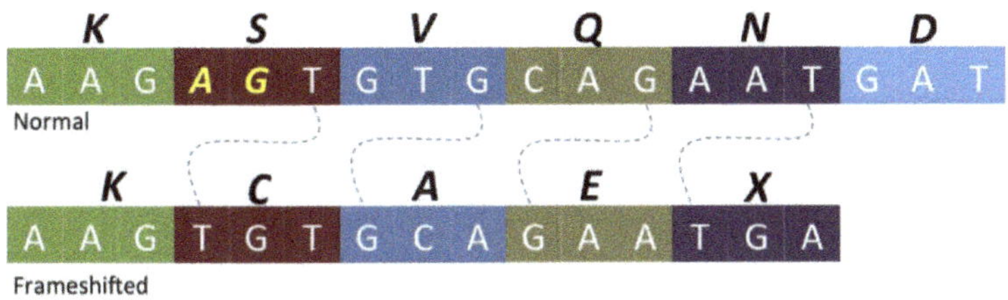

The normal gene is on top. The gene with the characters AG removed is at the bottom, as in Tara's genome. The amino acid specified by each triplet is indicated in bold above that triplet. The missing AG causes a major recipe change.

You cannot miss the dramatic change in *Tara*'s triplets above; they appear to be completely different once you go past the missing characters AG. The reason: triplet frames are now shifted on account of the two missing characters, as indicated by the dotted lines above. Hence the name *frameshift* for this type of recipe change. The amino acids generated are completely different as a result, starting from amino acid 711, where an *S* changes to a *C*. Thereafter, the recipe becomes completely different.

That's not all; something even more dramatic happens next. After amino acid 713, the recipe encounters a TGA triplet, which represents a *Stop* instruction marked by an *X*. That is the end of the recipe. The protein now has only 713 amino acids as opposed to 2871, a rather pale fraction of its normal self.

Had *Tara*'s recipe lost one full triplet instead of just two characters, the change in recipe would have been less dramatic. That triplet would have been lost but the rest of the triplets would have continued unaffected. Then her recipe would have been normal, but for the missing triplet. And unless that triplet was of particular importance, *Tara* would have continued leading a normal life. Unfortunately, losing characters in non-multiples of three has more drastic consequences, as was the case for *Tara*. Was this dramatic recipe change at the heart of her problems?

Rivets Come Loose?

Recall how muscle cells in the heart are glued together by rivets called desmosomes. Sarcomeres in muscle cells are then glued to these rivets by filaments made of desmin. This glue holds the entire ensemble together so it can withstand mechanical stress in the midst of energetic contraction and expansion. Here is a close-up of a desmosome.

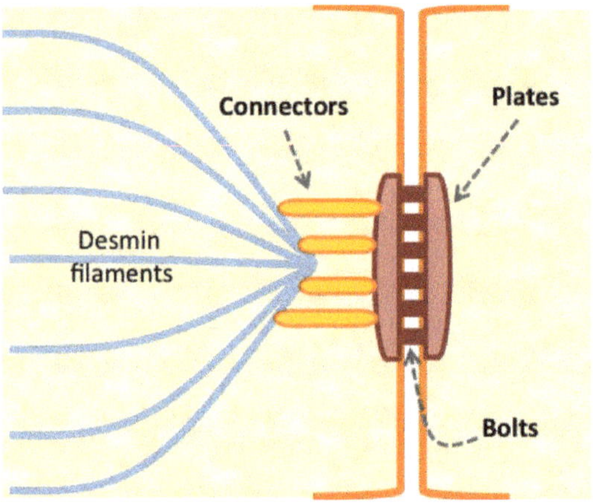

Desmosome plates, bolts and connectors.

As this picture shows, desmosomes have various components—plates, bolts and connectors. These components are made from proteins whose recipes are written in several genes. The recipe for the connectors is carried by a gene called *DSP*. This gene is of particular importance for this story, as it is the very same gene in which *Tara* had her two missing characters.

These missing characters forced a frameshift in one of *Tara*'s two *DSP* gene versions. The recipe in that version was truncated as a consequence. The connectors created from this recipe were therefore just about a quarter of their usual length. Such truncated connectors were probably incapable of anchoring desmin filaments to desmosome plates. The glue that provides a structural foundation for *Tara*'s heart was possibly weaker as a result.

Several questions arise now. Did the truncated connectors compromise the heart to such an extent that it could no longer bear the stress it was subjected to in its untiring quest to serve the body's need for blood and oxygen? Had it therefore yielded to the wear and tear of years and years of pumping blood? Is this what led to her heart failure at such a young age?

Possibly, though a niggling doubt remains. Fortunately, *Tara*'s second gene version remains unaffected, and continues to provide good connectors. Couldn't these good connectors hold the fort? Or were these inadequate when diluted with faulty connectors? We dig further for answers to these questions.

No Nonsense

Tara's faulty connectors, at just about a quarter of their usual length, were unlikely to tether desmin filaments to desmosome plates. Clearly, it would be unwise for cells to employ these faulty connectors; that would serve little purpose. Wouldn't it make sense to identify and weed out these faulty connectors instead? Indeed, cells have evolved mechanisms by which they can recognize and destroy connectors produced from truncated recipes (also called *nonsense* recipes, after which the destruction mechanism itself is called *Nonsense-mediated Decay*).

Isn't that magical? How would a cell know that a recipe has been truncated? After all, every recipe has a *Stop* triplet at its end. Imagine a cell as it executes the recipe for a particular gene. It encounters a *Stop*. Is this the normal intended *Stop*, or is it an abnormal, premature *Stop*? How would the cell differentiate between these two scenarios?

Some background first before we get to the answer. A gene could have many exons. Preceding the first of these exons is a stretch of text, called *UTR* (short for untranslated region, so called because it is not part of the gene recipe). Similarly, following the last exon lies another stretch of text, also called UTR. This UTR too is not part of the recipe; it simply follows the recipe, as shown below.

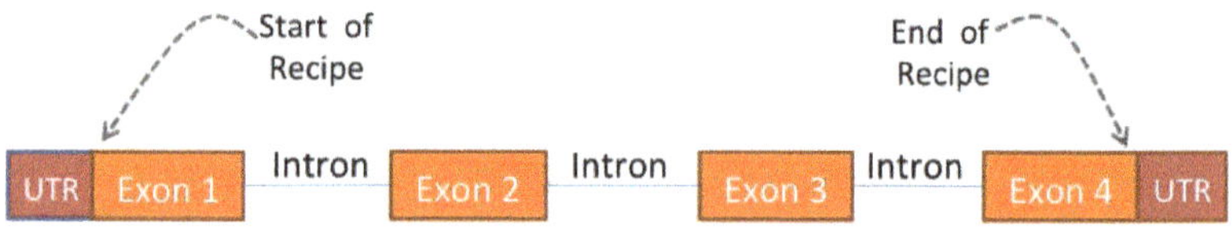

UTRs sandwiching the exons.

Starting with this picture, a walk-through of the recipe execution process helps answer our question. First, the entire stretch of the genome from UTR to UTR is copied into a template, which contains the two UTRs, all the exons, and all the intervening introns. Special-purpose molecules then excise out the introns so the exons now appear contiguously. These molecules are diligent though, so they leave markers at the ends of each exon to let everyone know where the exon boundaries were before they removed the introns. This is how the template appears now.

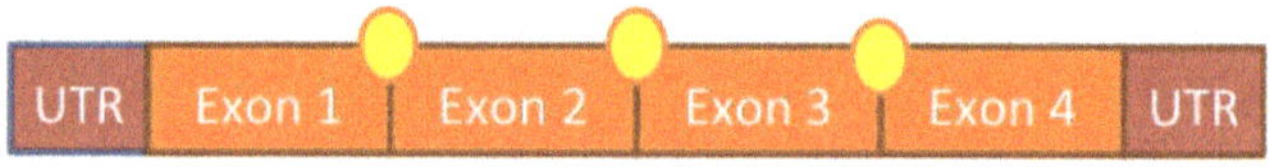

Introns gone but exon boundary markers, shown as circles, remain.

Next, the recipe is executed left to right by reading each successive triplet and appending the corresponding amino acid to the protein being manufactured. As soon as an exon boundary is crossed in this process, the marker at that boundary is removed. Recipe execution stops when a *Stop* triplet is encountered. This process can repeat many times, thus yielding many protein copies from the same template. So far, so good. But how does the cell know whether the *Stop* triplet encountered is the normal intended *Stop*, or an abnormal variant?

A simple check yields the answer now. If no exon boundary markers are left when the *Stop* is encountered, then all is well. On the other hand, if an exon boundary marker is still in place when a *Stop* is encountered, then this *Stop* is likely to be a premature one; the entire template is then destroyed so no recipe execution can happen from it.

Of course, not all premature truncations can be identified this way. For instance, truncations within the last exon are clearly spared, for all exon boundary markers have been removed by the time you get to the last exon. In reality, truncations in the penultimate exon within about 50 characters of its right boundary are also spared. The average length of an exon is only around 150 characters; this means that only truncations in the last 200 or so characters in the gene recipe are spared. All other premature truncations are typically identified and destroyed.

However, as with several processes in biology, this process is not perfect. There are rare exceptions—truncated recipes that escape nonsense-mediated decay and thus yield truncated proteins. We saw a possible example earlier in the context of our DCM patient *Nero*. Did *Tara*'s truncated recipe also escape nonsense-mediated decay, so her desmosomes received a mix of good and faulty connectors? Or was this truncated recipe eliminated as intended, leaving only half the number of connectors as compared to normal? Which of these two situations poses the greater danger?

Unfortunately, definitive answers can be obtained only through painstaking experiments which are not easy to perform for every patient. Instead, we have to rely on thought experiments and various other pieces of evidence to make an educated guess. Here is a simplistic argument though which suggests that a mix of good and faulty connectors poses the greater danger.

Suppose 100 good *DSP* connectors are needed to provide strong enough glue. And a normal individual actually generates, say, x connectors, where x is 100 or more (any leftover copies may be disposed of). In an individual with one of the two versions of the recipe truncated, there are $x/2$ good copies and $x/2$ truncated copies. If the truncated copies are destroyed by nonsense-mediated decay then only $x/2$ copies remain, so problems occur only if x is less than 200. But if the truncated copies are not destroyed by nonsense-mediated decay then problems occur no matter what. Why? For instance, suppose x were 400. Then there would be 200 good copies and 200 truncated copies. These would compete against each other for the 100 slots available. These 100 slots would then be filled in roughly equal proportion. So, half the slots would get truncated copies, compromising the glue considerably. Accordingly, the best case is that all truncated connectors are destroyed leaving only good connectors, but at half the normal amount.

What happens in this best case? Does it provide for a strong enough structural foundation for *Tara*'s heart? In which case, the missing characters in the *DSP* gene are not the cause of her problems. Or, alternatively, is this best case itself not good enough for *Tara*'s heart?

Of Mice

Answers to this question emerge from somewhat unexpected quarters: from experiments on other animals.

Mice can be created in a laboratory with a specific gene *knocked-out*, which means their genomes can be engineered so the recipe for that specific gene is disrupted and no protein comes from it. This engineering has indeed been accomplished with the *DSP* gene. And the results of this experiment provide us with interesting clues.

When both *DSP* gene versions were knocked-out, the effect was dramatic—the mice died while they were still tiny embryos.[30] Of course, desmin filaments could not connect to desmosome plates in these mice at all. More surprisingly, the number of desmosomes itself was far fewer than normal. When this knock-out effect was applied only to heart cells, leaving the other cells untouched, the mice did survive a bit longer, but they still died within a couple of weeks of birth.[31] Heart muscle cells were clearly disorganized in these mice.

All this when both gene versions of *DSP* were knocked-out.

What happened when only one of the two *DSP* gene versions was knocked-out, mimicking *Tara*'s condition? These mice were born normally—so one working version of the recipe seemed to suffice. But, wait! 20% of these mice died within six months of birth.[31] The typical lifespan of a mouse is about two years, so 20% of these mice lived less than a quarter of their typical span. Something was clearly amiss.

Further investigation showed a lot more was amiss. There were fewer connexins; remember, these transmit the heartbeat from one cell to another. So, heartbeat transmission was possibly slower than usual.[32] There were no structural changes or abnormal rhythms, initially, when the mice were two months old. However, when stressed by exercise, abnormal rhythms came to the fore. Over time, structural and rhythmic changes become more overt. The right and left ventricle chambers became enlarged. The left ventricle walls became thinner. Less blood was pumped out with each heartbeat. The heart attempted to compensate by increasing the heartbeat rate to abnormally high levels. This isn't very different from what was observed in *Tara*.

There was even more in store when hearts of these mice were observed under a microscope. Muscle cells in these mice were not organized as they should be. There was an excess of fat among these muscle cells.[31] And an excess of fibrous tissue, similar to what one sees in scars which form when cells die. These scars possibly interrupted the spread of electrical signal from cell to cell, thus leading to arrhythmia.

So, it appears that even the best case with all faulty connectors destroyed leaves the heart with a compromised structure. Good connectors at half their usual amounts can hold the fort for some time; but they inevitably give in to the long-term rigors of repeated pumping. Accelerated wear and tear then takes its toll. This seems to explain *Tara*'s condition.

This is evidence from mice though. Does that translate to humans? After all, humans are not mice.

.. and Men

Mice and humans do look very different to the untrained eye. Yet, we aren't distant relatives in a genomic sense. Our genomic text stretches actually resemble theirs for many important genes. Several genes whose knock-outs cause early death in mice also cause disease in humans. It is therefore reasonable to surmise that premature truncation in the *DSP* gene recipe is indeed the source of *Tara*'s problems. However, humans and mice are also different in many ways and we need substantiating evidence in humans to add to our conviction here.

Accordingly, we search scientific literature for patients with heart complications and with one of the two versions of *DSP* prematurely truncated. A number of such cases appear in the literature (for instance, one publication itself reports seven such individuals;[33] another reports two more such individuals[34]). That is certainly added evidence. But there is also a twist.

Consider an example: a 49-year-old woman with serious heart complications (abnormally fast heartbeat and thinning of muscle walls) has one version of the *DSP* recipe truncated after 1269 amino acids.[35] Her brother and son too have the same genomic variant. Their complications, albeit milder, are present nevertheless. So far, so good. Then comes the woman's 85-year-old father. He too has the variant. However, he shows no signs of any of the expected heart complications.

There are other such examples as well. For instance, a database of genomic variants from 60,000 presumed healthy individuals has 33 individuals with premature truncations, all in a single gene version each.[15] To summarize, it appears as if not everyone with one truncated *DSP* recipe suffers from a heart problem. Are we on the wrong track completely? Maybe this variant is not the cause of *Tara*'s problems at all.

Let's do a simple calculation to take stock. Among the 60,000 individuals mentioned above, all largely unrelated to each other, only 33 have a premature truncation. This is a frequency of roughly one in 2000, as far as unrelated, healthy individuals are concerned. Let us compare this with the corresponding frequency for patients with heart disease, as gleaned from two studies.

The first study[33] screened 100 families with ARVD (remember, a disease caused by problems in the desmosome) for variants in five key genes that yield various desmosome components. Seven premature truncations were found in this process. The second study[34] screened 121 DCM patients for variants in 46 genes. Two premature truncations were found in this case. So, among unrelated individuals with either ARVD or DCM, premature truncations appear with a frequency of roughly nine in 221.

Here are the two fractions we need to compare: one in 2000, versus nine in 221. They are not even close. In broad strokes, it appears that premature truncations in *DSP* are 75 times more likely to be present in individuals with ARVD or DCM than in healthy individuals.

A 75-fold greater likelihood makes a convincing case. Add to that what knock-out mice tell us. Taken together, it does appear that premature truncation in a single *DSP* version is the likely cause of *Tara*'s condition. No doubt, the effects of a single version truncation are not fully consistent. Not fully *penetrant* is the right term—not everyone with such a variant gets the disease. Studies indicate that 35% of the individuals with truncations in one version of any of the five desmosome-related genes show signs of disease by the time they are 40, and 60% show signs during their lifetime.[33]

Why do some individuals with truncation in one gene version show heart disease while the others don't? Possibly, some truncations escape nonsense-mediated decay while the others don't. And those which do lead to more severe disease. There is at least one example of a *DSP* truncation at amino acid 586 which escapes nonsense-mediated decay—it was found in 12 patients with heart disease, all from the same family.[36] *Tara*'s truncation at amino acid 713, is not far away. Maybe proteins created from this truncated recipe too escaped nonsense-mediated decay, at least to some extent, and compromised the glue even further.

Even in the above family with 12 patients, there appear to be four other individuals who likely have the same genomic variant as the 12 patients, but whose hearts function normally.[36] What causes this variation within the same family? Maybe half the normal amount of *DSP* protein pushes the heart close to the edge in its ability to withstand mechanical stresses, without tipping it over fully. Tipping over happens with time due to wear and tear, at rates that might vary from person to person. Indeed, *Tara*'s condition was much more advanced than *S*'s, even though both shared the same problematic variant.

Additional genomic variants in other heart-related genes could possibly modulate these rates and accelerate heart failure—there are indeed reports of variants in two distinct desmosome genes together increasing penetrance five-fold.[33] *Tara*'s genome doesn't appear to have any such additional variants. Could the *LAMA4* or *TTN* variants we kept aside play this abetting role? We just don't know.

Wrapping Up

When an illness of mysterious cause runs in a family, it leaves a sense of acute helplessness, a sense of not knowing whose turn it will be next and what they did to be subjected to this fate. Against this backdrop, a diagnosis is a relief. *Tara* and her family, through their difficult times, can take some consolation from knowing the likely cause of their affliction: a premature truncation in one of the two versions of the *DSP* gene.

An element of doubt continues to linger in this diagnosis however. *Tara*'s variant has never ever been seen in another patient. To be doubly sure, we submit *Tara*'s variant to a repository of variants called *ClinVar*.[37] Laboratories around the world submit variants from their patients to this repository. There is an outside chance that another person with the exact same variant would have their genome sequenced in one of these laboratories, and that variant would find its way into this repository eventually. For a while, there is no luck. However, two months later, another laboratory finds the exact same variant in a DCM patient. Like *Tara*, heart muscle walls in that patient too had grown weaker. That puts our diagnosis beyond doubt.

A *DSP* variant would conventionally lead to a diagnosis of *Arrhythmogenic Right Ventricular Dysplasia*—primarily, the right ventricle is affected. Muscle cells in this ventricle are progressively replaced by fat and fibrous tissue. The muscle wall thins and the cavity becomes larger. The left ventricle is also affected but only as disease progresses. In contrast, *Tara*'s condition was primarily left ventricular.

Interestingly, there is at least one reference in the literature to a truncating variant in *DSP* that affects the left ventricle primarily[36]—a reminder of how varied and unexpected the impact of recipe changes in this gene can be. *Tara*'s truncation, after 713 amino acids, is indeed somewhat similar to the truncation after 586 amino acids reported in that case. So, the right diagnosis for *Tara*'s condition would be *Arrhythmogenic Left Ventricular Dysplasia.*[36]

Where does this leave *Tara* and her family? *Tara* has an ICD implanted to counter arrhythmia. She has another implanted device, an *LVAD* or *Left Ventricle Assist Device*, which assists her left ventricle in pumping blood to her body. She awaits a heart transplant—that is the only cure known. Knowledge of the *DSP* variant does not offer her a cure today. Gene therapy of the type we saw in *Dia*'s vision loss story, while feasible in the eye, is not yet an option for the heart.

Tara's sister *S* also has an ICD implanted to counter her abnormal heart rhythms. She does not have heart failure yet—her left ventricle is not heavily compromised, so she has no overt symptoms. These could develop over time, so she needs to be monitored closely. It is known that stress-inducing exercise advances the age at which symptoms manifest in carriers of desmosomal variants, so she may need to restrict exercise.[38]

Tara's brother *B* and *S*'s young children show no abnormalities today. We do not know yet if they carry the *DSP* variant. If they indeed do and if behavior in mice is any indication, they will be at increased risk for cardiac arrhythmia when stressed, even if there are no apparent structural or heartbeat abnormalities. So, monitoring them regularly will be important.[32] On the other hand, if they do not carry the *DSP* variant, then all cause for concern goes away. They have the option of getting tested for this variant.

Our fundamental inability to repair damaged hearts to cure people like *Tara* remains though. Can we do better? Indeed, we have a source of inspiration: unlike human hearts, zebrafish hearts self-repair after an injury. Research into genes involved in this process might enable us to replicate this system in humans in the future[39]. So, some day, we might be able to learn from this little fish how to mend hearts with faulty connectors.

Story 4

The Mystery of the Eyes

Mira and *Sai* were healthy young parents, both in their 20s. Their second child *Som*, a boy, had an uneventful birth. Soon thereafter, repeated bouts of vomiting sounded an alarm. Investigations revealed abnormally thick muscle at the junction of his stomach and small intestine blocking the regular passage of food. By itself, this wasn't a major cause of concern—three in 1000 babies are so affected. A surgical procedure was performed to open the blocked passage. Vomiting stopped as a result, providing the necessary relief.

Unfortunately, additional complications soon came to the fore. Investigations determined that *Som* had *hypertension*, or high blood pressure. This was unusual, for hypertension usually develops with age. Its presence in babies is rare, though not unheard of. *Som*'s hypertension was of an even rarer type, for he suffered from both *systemic* and *pulmonary* hypertension. The former affects blood vessels that carry blood from the heart to much of the body, while the latter affects those that carry blood from the heart to the lung.

The root cause of this hypertension remained elusive in spite of much investigation. Abnormalities in the kidney can trick it into releasing enzymes that induce hypertension. A kidney scan on *Som* didn't uncover any such abnormality. Defects in heart structure can also cause hypertension; again, a scan of *Som*'s heart failed to yield any clue. Abnormal hormone levels can also cause hypertension; yet again, blood tests to check for these came back normal.

Som's complications did not end there. When he was ten months old, he got a lung infection—an attack of *pneumonia*. He recovered from this. But then, rather abruptly, things took a turn for the worse. When he was just 17 months old, *Som* developed breathing difficulties and passed away suddenly. This was, of course, a tragic fate for *Mira* and *Sai*. Even more so because their first child *Rom*, also a boy, had met an identical fate—he passed away due to breathing difficulties when he was just 13 months old.

The loss of their two children, no doubt, left *Mira* and *Sai* distraught. Their children's symptoms did not fit a textbook mold, making diagnosis difficult even for the best of physicians and geneticists. The inability of doctors to explain this mysterious affliction left them helpless. There was no history of any such early deaths in the family to provide any clue. *Mira* and *Sai* were still young enough to have more children. But would they be able to bear this emotional trauma yet again?

Som and *Rom* both had another strikingly unusual feature which we haven't talked about yet; can you spot this feature in the picture below?

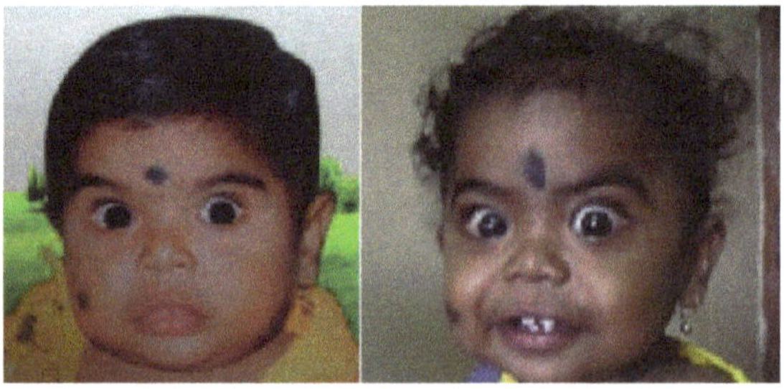

Prominent eyes in Som and Rom.

A trained eye can spot subtle disproportions in faces—the eyes being slightly further apart, or the forehead being slightly more prominent, or the mouth being slightly larger. One doesn't need a trained eye though to spot the very prominent, bulging eyes in the picture above. Aren't they unusually prominent?

A trained eye might be able to go further, as the following

experiment will convince you. Observe faces around you and note the distance from the base of the nose to the mouth, and the distance from the mouth to the bottom of the chin. Which distance is larger in the faces you observe? Do the same proportions hold for the faces above? Or are *Som*'s and *Rom*'s chins disproportionately small?

Could these prominent eyes, small chins, and full cheeks in *Som* and *Rom* hold a clue to the root cause of their tragic fate?

Where Do We Look?

Since both boys suffered from similar complications, a shared genomic character is the likely cause of their tragic fate. Since *Rom* had passed away much earlier, only the genomes of *Mira*, *Sai*, and *Som* were available for our investigation. Our challenge is to identify the offending genomic character by sequencing these genomes.

As in previous stories, we focus only on genes—recipes embedded within the genome. We extract and sequence exons from all 20,000 genes. This yields tens of millions of reads, or tiny shreds of the genome. After we assemble the jigsaw puzzle posed by these reads, we obtain a list of variant characters—at least one of *Mira*, *Sai*, or *Som* differs from the reference sequence in each of these characters. The reference sequence represents a healthy individual. So, the key to the mystery in this family lies, possibly, in one or two of these variant characters, or just variants, for short. As before, our main problem is that there are tens of thousands of these variants. How do we identify the needle in this vast haystack?

We start by focussing on genes known to cause symptoms like those experienced by *Som* and *Rom*. Unlike in previous stories, where symptoms were restricted to a single organ like the eye or the heart, the symptoms here are more wide-ranging. There are changes in the facial structure (prominent, bulging eyes), possibly in the blood vessels (hypertension), possibly in the lungs (respiratory difficulties), possibly in the heart (sudden death), and possibly in the stomach and intestine (thickening of the stomach muscle). The object of our hunt seems to have its tentacles spread far and wide. Our list spanning this gamut of symptoms comprises 500 or so genes—too many to sift through one by one.

We now turn our attention to variants in these genes, looking for anything suspicious that *Som* and *Rom* inherited from their parents. A caveat before we rush ahead—not all genomic characters are inherited from parents. A few are *de novo* variants, introduced by error when a parent's genome is copied to create the child's genome. Each of us may have few tens of such variants. However, it is very unlikely that such de novo variants appeared independently in both *Som* and *Rom* resulting in the same tragic fate. So, our focus remains on variants that the boys inherited from their parents. In particular, we are interested in variants that *both* boys must have inherited. Had we access to genomes from both boys, we could have reduced the size of our haystack substantially by identifying such shared variants. Unfortunately, *Rom*'s genome is not available. So, we must work harder to find our quarry based on evidence from the genomes of *Mira*, *Sai*, and *Som* alone.

This poses a natural question: if *Som* inherited a lethal variant in one of these 500 or so genes from his parents, how is it that the parents remain healthy?

Recessive, Dominant? Or X?

Of course, the answer to this question goes back to Mendel's pea plant experiments, which we encountered in *Dia*'s story of vision loss. Remember, each of us has two versions of most genes. We inherit one version from our father, and the other from our mother. Occasionally, one of these versions develops a fault. The ill-effects of this faulty version manifest only when it is not dominated by the good version. In our family, the version with the lethal variant is probably dominated by its good counterpart. So, the parents are protected because they have one good version each. *Som* and *Rom* probably inherited the lethal version from both parents, thus allowing it to unleash its ill-effects unchecked.

But what are the odds of *both* parents carrying a lethal version?

Typically, minuscule. Our family is unusual though because *Mira* and *Sai* are genetically related. In fact, they are uncle and niece, respectively—not an uncommon occurrence in parts of South India. Because of this consanguinity, the chances of such a freak occurrence are far more pronounced.

Of course, we must remember that *Mira* and *Sai* need not carry the same lethal variant; they could well carry two different lethal variants. Remember, *Dia*'s vision problems in the second story presented a similar picture; her parents carried two different problematic variants in the *ABCA4* gene, and, unfortunately, *Dia* had inherited both. The same could be true here. *Som* may have inherited one lethal variant from each parent, an event with a 25% chance. The chances of *Som* and *Rom* both doing so would be one in 16, or about 6.25%—small, but not improbably so.

There is another scenario as well—what if the offending variant is on the X chromosome? Remember, we have 23 pairs of chromosomes, and the 23rd pair comprising the X and Y chromosomes determines sex. Females have two versions of X and none of Y, while males have one of each. *Som* and *Rom* were both boys; they had only one version of X each, which they inherited from their mother *Mira*. Their father *Sai* had no say in the matter.

Mira herself has two versions of X. If the lethal variant is present in only one of these two versions, she is protected by the other version. But if *Som* had inherited this lethal variant in his sole version of X, then he would be squarely exposed to its ill-effects in the absence of a second version. The chances of this unfortunate happenstance in both boys, each at 50%, would be one in four, or 25%—not so small at all.

Accordingly, we instruct our computer to look for variants in our shortlist of about 500 genes which satisfy one of the patterns shown below.

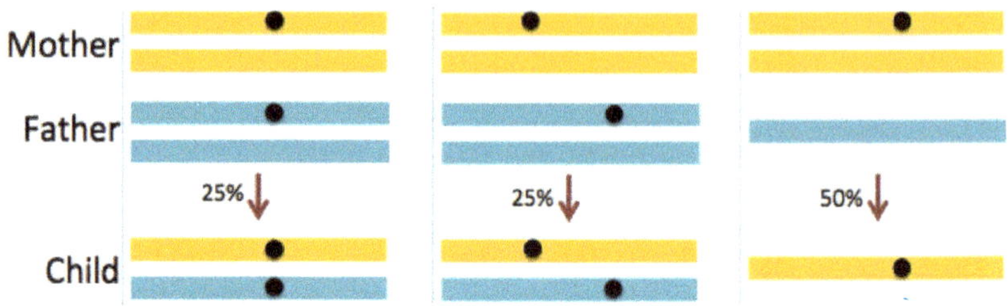

Possible inheritance scenarios. The last scenario involves the X chromosome.

As in previous stories, we look for variants that have a significant impact on the gene recipe, ignoring others which leave the recipe untouched. We specifically look for missense variants which modify one amino acid into another, nonsense variants which create a premature *Stop* instruction leading to a truncated recipe, and frameshift variants, which change the sequence of amino acids thereafter, almost always culminating in a truncated recipe as well. Of these, we look only for *rare* variants which do not appear in many people, because common variants are unlikely to cause the rather rare condition we saw in *Som* and *Rom*.

Our computer wades through the tens of thousands of variants looking for those that match the above description, and obligingly produces just a handful. As we pore over these candidates, we eliminate all but two for various reasons, leaving us with just two suspects. Meet the genes: *Fibulin 5* or *FBLN5*, and *Filamin A* or *FLNA*. What do these genes do, and is there sufficient evidence to bring a watertight case against one or both of them?

Loose Skin and False Starts

Scientific literature does mention variants in the *FBLN5* gene that cause pulmonary hypertension, repeated respiratory infections, and death due to respiratory failure in the first two years of life.[40] These were all features which *Som* shared, making us sit up and take note.

But the geneticist who had treated *Som* would have none of it. Just the mere mention of the *FBLN5* gene meets with categorical disapproval: "Certainly not! These boys had no *Cutis Laxa*", she said.

Indeed, cutis laxa, which translates to loose skin, is the hallmark of variants in *FBLN5*. Typically, our skin is elastic; you pinch it, and it pulls right back into its place. Not when one has cutis laxa though; the skin becomes loose and saggy, like a piece of cloth. If indeed a variant in this gene were the culprit, we would expect to see signs of cutis laxa. However, both *Som* and *Rom* had perfectly normal, elastic skin. Shouldn't this rule out an *FBLN5* variant as the likely cause?

A careful investigation that leaves no stone unturned must ask the following question at this point: could the *FBLN5* variant in *Som* indeed cause all his other problems, yet leave his skin untouched? A glance at *Som*'s variant tells us that there might well be more to it than meets the eye.

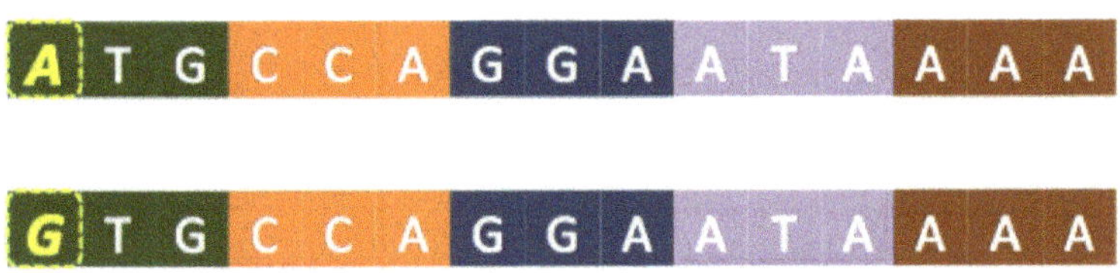

The FBLN5 reference sequence on top, and Som's FBLN5 sequence at the bottom. The start triplet ATG changes to GTG in both versions of the FBLN5 recipe in Som.

Remember, gene recipes are organized as character triplets. The end of the recipe is signaled by a special *Stop* triplet—either TGA, TAA, or TAG; we encountered this triplet multiple times in previous stories. Along similar lines, the recipe begins with a special *Start* triplet, which we haven't encountered yet. This start triplet comprises the characters ATG. While ATG triplets can occur anywhere in the recipe, and they always specify the amino acid *M*, only the very first of these ATGs also doubles up as a start indicator. Recipe execution begins from this ATG. In *Som*, this first ATG had rather interestingly become a GTG in both gene versions.

GTG is probably too weak a start indicator by all accounts.[41] So, both versions of the *FBLN5* recipe in *Som* had lost their start markers. The recipe execution machinery no longer knew where to begin. Did it then entirely fail to execute the recipe and manufacture the *FBLN5* protein? If so, then *Som* would have had loose skin, which he clearly didn't. Something must have come to his rescue. Possibly, the recipe execution machinery continued to search further into the recipe for an alternative start indicator, blissfully unaware that the original object of its quest was no longer to be found.

There are many more ATG triplets in the *FBLN5* recipe. Couldn't the very next such ATG serve as the new start point for the recipe? Possibly not, for a certain character pattern in its neighborhood is needed for it to do so. This pattern of characters is called the *Kozak* sequence,[42] shown below.

The Kozak pattern around an ATG start triplet, showing the most likely characters at each position. The starred positions are relatively more important. A/G indicates both characters are possible; the second character is shown in smaller font if it is less likely.

The Kozak pattern is not an absolute requirement—it demands only approximate adherence. Of the various positions in this pattern, a match at the starred positions is more important. In particular, the first starred position has a dominating effect.[42] A match at this position allows some leeway for a mismatch at the second starred position. The more the pattern around an ATG matches the Kozak sequence, the greater the chance of recipe execution starting from that triplet. Of course, there could be several ATG triplets in the recipe all vying to be starting points for recipe execution. Indeed, the next six ATGs in *FBLN5* are shown below.

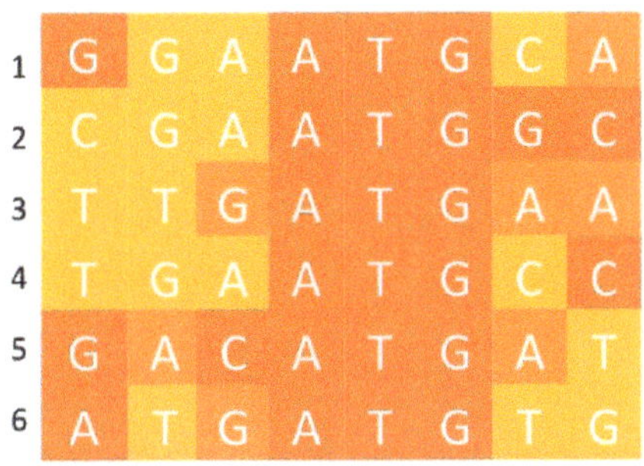

Six candidate ATGs along with characters in their vicinity in each case. Dark shades indicate matches to the Kozak pattern, light shades indicate mismatches, and intermediate shades depict matching to the second-most-likely character.

Which of these ATGs emerges the winner? Since recipe execution scans the recipe from left to right, an earlier ATG stands a better chance, unless, of course, it matches the Kozak pattern poorly. A glance at the picture above indicates that this tussle between being the first in line and being the best match to the Kozak pattern is very likely won by the ATG in row 5. Several methods to quantify this match have been invented by researchers. According to one of these,[43] the ATG in row 5 presents a better match to the Kozak pattern than do the actual start triplets in as many as 25% of the 20,000 genes. And only 17% of ATG occurrences known to be NOT valid start triplets have better matches. The ATG in row 5 does have the makings of a likely winner.

This victorious ATG appears 168 characters into the recipe from the original start point, well past the first four ATGs. However, unlike the first four ATGs, this one presents a saving grace: 168 is divisible by 3, and therefore, the recipe, despite losing $168/3 = 56$

amino acids, stays in frame thereafter, thus retaining much of its character. If any of the first four ATGs had emerged victorious, the recipe would have gone out of frame yielding a completely new set of triplets, as had happened to *Tara*'s *DSP* gene in the previous story. So, *Som*'s *FBLN5* recipe is likely shorter by 56 amino acids, but otherwise intact.

Could the loss of these 56 leading amino acids explain *Som*'s condition while simultaneously explaining the lack of cutis laxa? The picture below suggests an answer by depicting where variants known to cause cutis laxa in patients lie in the recipe.

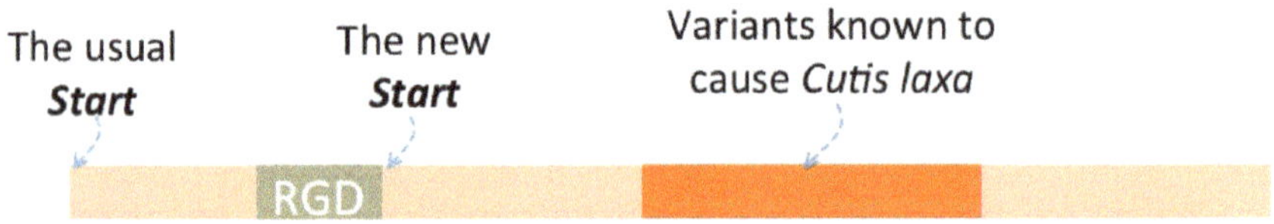

The FBLN5 recipe, with the new starting position and the region with missense variants causing cutis laxa marked. The last three amino acids which precede this new starting position carry the short forms R, G and D, respectively.

Only a handful of missense variants in this recipe are known to cause cutis laxa today. Remember, missense variants change one amino acid in the recipe into another. Many more variants will no doubt be discovered in the future. However, for the moment, it appears as if all missense variants causing cutis laxa lie in a region that stays intact even when the first 56 amino acids are lost. In other words, the relevant parts of the recipe are preserved in *Som*. That would explain his lack of cutis laxa. So far, so good. This leads us now to our key question: could the loss of the first 56 amino acids somehow cause hypertension, prominent eyes and early death in *Som*, while steering clear of cutis laxa?

A brief digression first. The elastic nature of skin comes from elastic fibers that are found outside the cells in our skin. These fibers are also found in other organs where elasticity is key, like the blood vessels and the lungs. Often, we have to rely on our cousins, the mice, to tell us what impact a particular gene might have on these fibers. Indeed, when mice are engineered so both versions of the *FBLN5* recipe are crippled, or knocked out, these fibers tend to be highly disorganized, leading to inelastic skin, and lung and blood vessel complications.[44] Experiments attempting to explain this phenomenon show how *FBLN5* might help organize these elastic fibers by anchoring them to cells as shown in this picture.

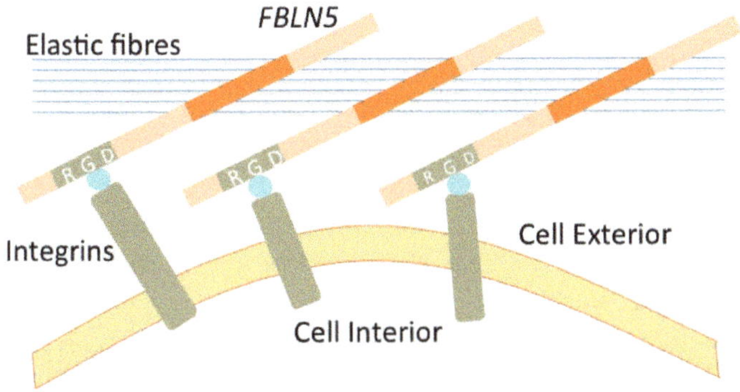

FBLN5 proteins anchoring elastic fibers to cells.

As you can see, attaching to these elastic fibers is the shaded/colored part of *FBLN5*, which carries missense variants that cause cutis laxa. In contrast, the part comprising amino acids *R*, *G* and *D* attaches to proteins called integrins, which appear on the surface of skins cells.[44] By attaching at both ends, *FBLN5* proteins anchor elastic fibers to skins cells. When the protein loses its first 56 amino acids, the end that attaches to the elastic fibers is unaffected, but the R–G–D section is lost. What happens then?

Maybe elastic fibers are no longer anchored to skin cells as they should be, thus reducing elasticity. In the blood vessels, this might mean stiffer walls, leading to increased blood pressure, which we saw in *Som*. In the lungs, this might mean less elastic air sacs which cannot expand and contract quite so fluently with each breath, possibly leading to the respiratory distress seen in *Som*. In the skin, this might mean cutis laxa. But *Som* did not have cutis laxa. So, maybe, just maybe, elasticity in the skin is not reduced quite to the extent needed for cutis laxa?

This might well be wishful thinking, but a careful investigation needs to leave no stone unturned. Which brings us to our main question: could the loss of the R–G–D section lead to reduction in elasticity in the blood vessels and in the lungs, but not in the skin?

One way, and possibly the only way, to answer this question definitively, is by experiment. With bated breath, we search for experiments that confirm the preferential reduction in elasticity in the blood vessels and in the lungs, relative to the skin. Our quest leads to a relevant experiment, where researchers created mice with the amino acid *D* in R–G–D replaced by another amino acid *E*, in both gene versions. The new combination no longer attaches to integrins on the cell surface and in that sense mimics the removal of the R–G–D section in its entirety. Do these mice show reduction in elasticity in the blood vessels and in the lungs, but not in the skin?

The answer is disappointing. These R–G–E mice show no abnormalities at all—not in the lungs, not in the blood vessels, and not in the skin.[45] We don't know why; possibly, other genes take over where *FBLN5* fails, a reminder of the redundancy that protects us. There is a saving grace, however: we might have an explanation of why *Som* showed no cutis laxa in spite of losing the start triplet. A new start salvaged the situation. And if behavior in mice is any indication, this salvage might have been quite effective.

This means that we are at a dead end with *FBLN5*. Dead ends in detective stories are not uncommon. No reason for despair though, we still have one more lead to pursue.

The Master Orchestrator

As we survey literature on our last lead, *FLNA*, what emerges is a picture of awe. Unlike genes we have encountered so far that play rather localized roles in one organ or the other, this gene seems to have its influence spread far and wide—the brain, the heart, the blood vessels, the lungs, the bones, and even the intestines. How does it spread its net so far and wide? A brief digression is needed to set the stage for the answer.

Imagine a single cell dividing and growing into a complete organism. Many different organs have to be created. Each organ has a specific shape and structure. For instance, blood vessels are long and tubular, the lungs are like balloons, the heart is a four-chambered thick-walled structure, and so on. As cells divide, more cells become available for the creation of these varied structures. These cells must then reorganize themselves. But, how? An analogy may be helpful at this point.

Imagine you were given a collection of bricks and asked to build a house. You would have to move bricks around to create various parts of the house. Further, you would have to cut bricks into varied shapes, as dictated by the geometry of the house. Similarly, cells have to be moved around and also morphed into various shapes. What moves them around and morphs them into desired shapes?

Shapes? Do cells have definite shapes? Aren't they simply like squishy round grapes? Not really; they do have shapes that are definite. Not only that, a cell can change from one definite shape to another, very deliberately. This ability of a cell to change its shape allows collections of cells to then morph themselves into specific structures as below.

A row of cells transforms itself into a collective circular shape as the upper ends expand and the lower ends contract.

For instance, imagine a row of cells as in the picture above, and suppose these need to be sculpted into a circular form. Assume these cells are glued firmly together. Each cell then modifies its shape by expanding the upper end and contracting the opposite lower end. The net result: the entire collection twists into a circular form.

This ability to change shape deliberately also enables a cell to move. It simply forms a protuberance that stretches forward in the direction of motion at the leading edge. Simultaneously, it retracts the opposite end at the trailing edge, much like an earthworm. Together, movement and shape changes sculpt the various organs in our body.

So, what gives shape to a cell? Inside the cell, particularly along the periphery, is a region called the cortex. It comprises a framework of crisscrossing filaments (similar to ropes), as shown in the picture below.

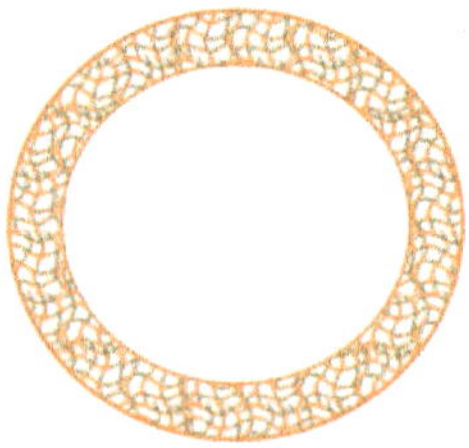

The cell cortex gives shape to the cell.

This framework is called the *actin cytoskeleton*, a skeleton, but of the cell. In the previous story, we encountered a network of *desmin* filaments; these are thicker than actin filaments, but found only in specific organs which encounter greater mechanical stress, like the heart. The actin cytoskeleton is found in most cells. Here is a close-up of the cytoskeleton.

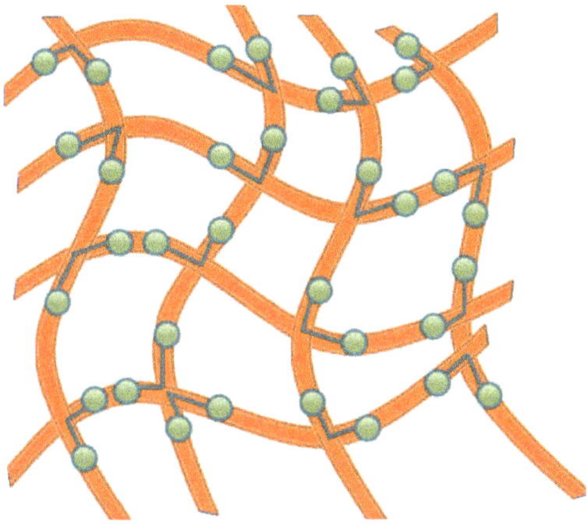

Crisscrossing actin filaments forming a three-dimensional cytoskeleton. Also shown are proteins that hold crisscrossing filaments together.

The filaments in the cytoskeleton need to be held together strongly for the cell to retain a specific shape. Several proteins aid in this task. Some connect parallel filaments together into bundles for greater strength. Others form connections between crisscrossing filaments, creating a three-dimensional network of filaments. The gene recipe of our prime suspect, *FLNA*, yields a protein that is tailor-made for the latter purpose.

To appreciate *FLNA*'s role, imagine a person at each filament junction holding a pair of crisscrossing filaments firmly, one in each hand, as depicted by the circles in the picture above. By adjusting these filaments in different parts of the cell in different ways, cells can now change their own shapes. For instance, a row of cells can morph into a curved shape if the filaments at one end are tightened while those at the other are loosened. Thus, the *FLNA* protein influences cell shape and cell movement, and therefore the formation of various organs.

Organ formation, though, is a complex task: various different shapes and movement strategies have to be deployed at different points of time. This programming requires chemical cues to be converted to shape and movement responses and vice versa. For instance, to enable a cell to move, the protein which serves as glue between adjacent cells needs to be destroyed. And once sufficient movement has happened, this protein needs to be regenerated. These mechanical and chemical changes go hand in hand, and the conversion from one to the other is also influenced by the *FLNA* protein.[46] Thus, by influencing cell shape, cell movement and the conversion between mechanical cues and chemical signals, *FLNA* plays the role of a master orchestrator in the development of various organs. Did a quirk in its recipe in *Som* throw this master orchestrator off balance?

Altered Recipes in *FLNA*

Studies on patients in whom the *FLNA* recipe is altered tell the sordid tale of its far-reaching impact. And as we listen to these stories, the picture turns from one of awe to one of horror. If there ever was a gene with an extensive criminal record, this might be the one.

The impact of *FLNA* is highly gender-dependent: on females it is morbid, but on males it is just brutally gruesome. This discrimination is a consequence of *FLNA*'s location in the genome—it appears on the X chromosome. Males have only one version of this chromosome, so there is no cover. The impact of a problematic variant can then be lethal. Indeed, literature is rife with instances where male babies die while still in the womb or shortly thereafter, with various horrifying malformations. Females can tolerate one version becoming dysfunctional because the other version can hold the fort. But *FLNA* is so potent that even one version becoming dysfunctional sometimes causes severe issues. The only saving grace though: immediate fatalities are less common.

One set of variants in *FLNA* causes what is called *Periventricular Heterotopia.*[47] Ventricles here are four fluid-filled cavities at the center of the brain, and not the ventricles in the heart which we encountered earlier in *Tara*'s story of heart failure. When the brain is forming early in life, neurons first form in the region around these ventricles, and then move outward to form the exterior surface of the brain. If this neuronal movement does not happen correctly, then neurons clump together near the ventricles instead, and hence the nomenclature Periventricular Heterotopia—periventricular means near the ventricles, and heterotopia means appearance at an unexpected position.

An interesting fact about the variants in *FLNA* which cause Periventricular Heterotopia is that several of these are nonsense and frameshift variants. Remember these are variants which truncate the gene recipe prematurely, rendering it non-functional. Most of these variants do not allow males to survive, so symptoms have been studied primarily in living females. There are a few exceptions—some missense variants which compromise the gene in a more minor way, allowing males to survive.

Though neuronal mislocation is the hallmark of nonsense and frameshift variants in *FLNA*, its impact on the brain is more muted than expected. In fact, it is often mild or even silent, detectable only on an MRI scan. That is, until the affected person starts experiencing seizures, or sudden surges of abnormal brain activity, starting in the teens or the twenties. Mental disability also happens less often than expected. The more serious symptoms occur in other organs.

In some individuals, heart valves and blood vessels close to the heart are malformed. In others, intestines are structurally affected and may lead to constipation. Some individuals have problems with blood clotting. A few might even develop life-threatening complications. For instance, one report[48] describes four girls with progressive respiratory failure in the first four months after birth. They also had abnormal crossflow of blood between two major blood vessels, as well as pulmonary hypertension. All the four girls needed lung transplantation for survival. This is indeed quite far-reaching, that too in spite of one gene version being fully normal—yet again, a reminder of the potency of *FLNA*.

There are other missense variants in *FLNA* whose primary impact lies not in the misplacement of neurons, but in changes to facial and bone structure. These variants seem to actually enhance the gene's ability to function.[49] This is another testament to how critically poised *FLNA*'s role is in the development of various organs—both increased as well as decreased potency leads to problems.

The abnormalities caused by these variants come in various shades and are accordingly classified into a number of complicated disease names.

Otopalatodigital syndrome type I (OPD1), *Otopalatodigital syndrome type II (OPD2)*, *Frontometaphyseal dysplasia (FMD)*, *Melnick-Needles syndrome (MNS)* and *Terminal osseous dysplasia (TOD)*.

In these diseases, the long bones of the legs are often bowed, the rib cage is smaller and weaker, and fingers and toes are often malformed. There are other far-reaching complications as well: deafness, and defects in the heart, brain, intestine, and kidneys. In addition, facial structure shows some unusual patterns.

The study of facial proportions becomes important in this context. Indeed, there is a whole branch of medicine dedicated to this theme—specialists trained in this area can spot subtle facial abnormalities, or dysmorphologies, as they are called. For instance, the bony ridge of the forehead, just above the eyebrows, may be unusually prominent (which is the norm in chimpanzees and gorillas, less so in humans). The eyes may be set further apart than usual. The distance from the base of the nose to the mouth, which ought to be less than the distance from the mouth to the bottom of the chin, might go the other way. It takes a trained eye to spot these subtleties though.

The impact of these variants, which primarily cause skeletal and facial abnormalities, is quite varied though. Males born with *OPD1* and *FMD* do survive with relatively mild symptoms. On the other hand, those with *OPD2* and *MNS* often die before birth. Even those fortunate enough to be born, die early in life due to respiratory failure. On occasion, they do survive beyond their second year, but only with intensive treatment.[50] Note that *death due to respiratory failure* appears to be a recurring theme in *FLNA*'s record book.

For females, as expected, these variants are typically not life-threatening. However, they do show a wide range of expressivity. Some females appear completely normal—sub-clinical is the term; there are no apparent abnormalities, though detailed tests might unearth subtle internal issues. Others show as much severity as males do. What causes this variability is not quite well-established yet.

So, that is *FLNA*'s rather notorious track record. Where do *Som* and *Rom*'s symptoms fit in all this? A close-up of the various sections of the *FLNA* recipe is all that remains for us to set the stage for this investigation.

A Close-up of *FLNA*

Our master orchestrator, *FLNA*, prefers not to work in isolation. Rather, two *FLNA* protein molecules join together at the hip and work in tandem. The section labeled **C** at the very end of the recipe is where they join to make a pair, as shown in the picture below.

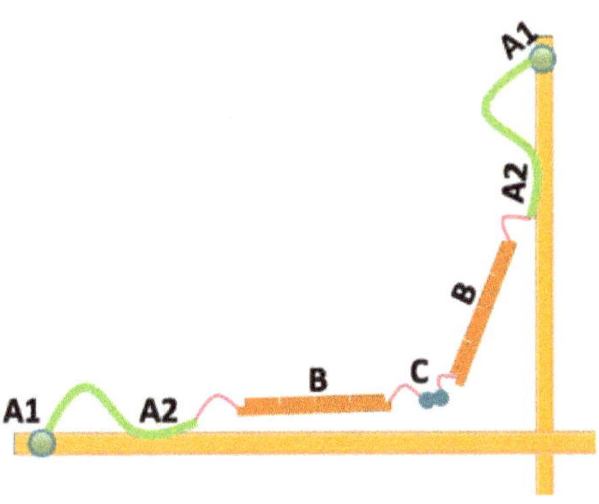

A pair of FLNA proteins, joined together at **C**, *each attached to a distinct filament of the cytoskeleton.*

This pairing is critical—when genomic variants in this section prevent pairing, things no longer function as intended. Once the pair forms, each member of the pair grasps a distinct crisscrossing filament of the cytoskeleton. The hands that grasp these filaments come from two different sections of the recipe, called **A1** and **A2**, as shown in this picture. The strength with which each member of the pair grasps a filament of the cytoskeleton is critical. Even subtle increases or decreases in this strength on account of genomic variation make *FLNA* behave differently, sometimes with dramatic effect.

In between the **A2** and the **C** sections of the recipe is a long section labeled **B**. The recipe in this section yields a special structure that unfolds as cytoskeletal filaments are stretched by mechanical forces acting upon the cell. In this unfolded state, section **B** interacts with a large number of other proteins and thereby initiates various chemical signals[46] with important cascading effects. This conversion of mechanical signals to chemical signals through the interaction between **B** and various other proteins is what makes *FLNA* the master orchestrator.

And this leads us to the burning issue: is *FLNA* the object of our hunt?

Is it *FLNA*?

The *FLNA* variant that we found in *Som* is a missense variant. *FLNA* lies on the X chromosome, and *Som* had only one version of this chromosome. This lone version has the variant. *Som*'s mother *Mira* too has this variant, but in exactly one of her two gene versions; possibly, the other gene version protects her. *Som*'s father *Sai* does not have this variant. We don't know whether *Rom*, *Som*'s brother, had this variant since he passed away much earlier. In any case, how do we know if this variant is a mere onlooker, or indeed the root cause of our family's troubles?

We examine the brothers' symptoms and compare these with what *FLNA* variants typically do. Remember, one hallmark condition was misplaced neurons. Did *Som* and *Rom* have these? Unfortunately, both boys had passed away well before *FLNA* entered the picture as a candidate; so neither was checked for these misplaced neurons.

Mira too had the variant in one gene version, so she might also have misplaced neurons. This would cause seizures—sudden surges of abnormal brain activity—starting in the teens or twenties. *Mira* was indeed in her twenties, but had experienced no such episodes. She was in her early twenties though, so these episodes might yet arise. However, her complete normalcy made this a somewhat slim possibility.

Going beyond misplaced neurons and looking for other related signs, no one in this family seemed to show any heart abnormalities, severe constipation or problems in blood clotting. So, *Som*'s missense variant seems not to fit the bill, at least as far as variants in *FLNA* causing misplaced neurons go.

How about the other category of variants in *FLNA*—those whose hallmark effects are on facial and skeletal structure. The boys in our case did have some very distinctive facial features. Do these fit a profile typical of *FLNA* variants? Let us remind ourselves with a picture of a girl with *MNS*, one of the syndromes caused by missense variants in *FLNA*. What in this picture strikes you as distinctive?

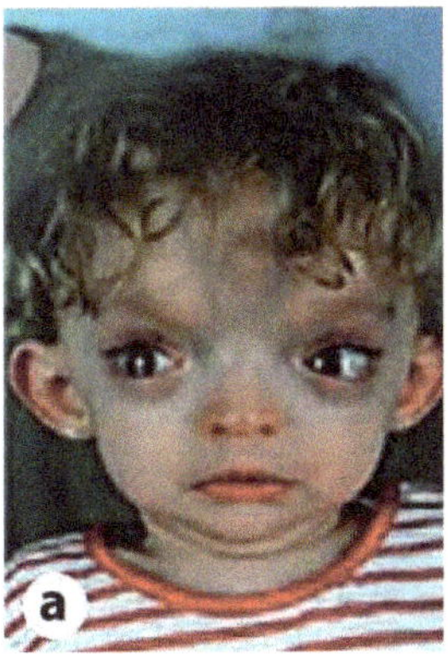

Typical facial structure in Melnick-Needles syndrome (MNS). Picture reproduced with permission.[51]

An expert on facial dysmorphology will not miss the prominent eyes, the full cheeks, and the small lower jaw. These are hallmark *MNS* features. The boys in this story too had prominent eyes. Very prominent actually, almost bulging from their sockets. Fairly full cheeks. And small lower jaws. Would this suggest that the boys in our case had *MNS*?

The clinician treating this family is not convinced. These classic facial features of *MNS* are typically found in females. Females do show prominent eyes, full cheeks and small lower jaws. While many live full lives, some females die of respiratory failure, typically in the second or third decade of life. Frequent respiratory infections occur. Pulmonary hypertension, or high blood pressure in the blood vessels linking the heart and the lung, is also known to occur.[52] *Som* and *Rom* clearly died of respiratory failure. *Som* did have one reported incident of pneumonia when he was ten months old. Both also showed pulmonary hypertension. The one catch: *Som* and *Rom* were boys, not girls. Boys with *MNS* are more likely to pass away while still in the womb, or soon after birth, and also suffer from serious skeletal deformities.

This gender riddle initiates another exploration of scientific literature. Indeed, we find most references to *MNS* in scientific literature are to affected females. Male children born to these affected females show severe skeletal and facial malformations and die of respiratory failure very early. So, living males are rare and have not been studied much. However, scientific literature points out a striking fact: there are two distinct types of male *MNS* patients.[53] The first group are born to affected mothers; they show severe disease and die very early. There is a second, rarer, group comprising males whose mothers are unaffected; these males show a milder form of the disease, similar to what affected females show. Maybe the boys in our family fall into the latter group.

What might cause this dual behavior? Maybe the specific variant within the *FLNA* gene? Some variants might cause moderate effects in females but very severe effects in males. Others might leave females unaffected while causing less severe effects in males. What does the specific variant we found in *Som* and his mother *Mira* tell us?

The Fertile Desert

The *FLNA* variant that *Som* and *Mira* both possess is a missense variant called *D1970N*—amino acid *D* at position 1970 in the recipe has been modified to amino acid *N*. *D* is short for Aspartic Acid and *N* for Asparagine. As you might guess, the latter gets its name because it was first isolated from asparagus juice. *Som* has it in his sole version of the *FLNA* recipe on his X chromosome, while *Mira* has it in just one of her two versions of the recipe.

We know that this variant has never been seen previously. Tens of thousands of people whose genomes have been sequenced all report a *D* at this position. So, there is absolutely no history to go by—we have to make indirect inferences. Nearby variants in the *FLNA* recipe found in other patients might yield a clue to the behavior of this variant. So, we draw the following picture which depicts most known problematic variants in *FLNA*.

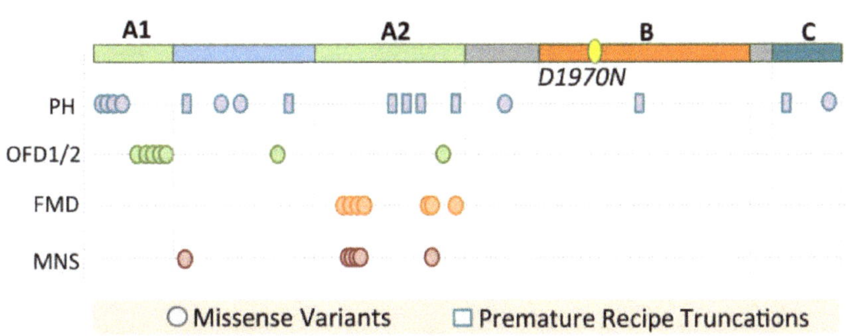

Disease-causing variants in FLNA found in patients with various diseases, as reported in scientific literature. All variants causing the same disease appear on the same horizontal line. PH stands for Periventricular Heterotopia, MNS for Melnick-Needles syndrome etc. D1970N appears as a yellow dot.

What does this picture scream out loud? Square markers, or premature recipe truncations, cause misplaced neurons, the hallmark of Periventricular Heterotopia. In contrast, round markers, or missense variants, appear in all diseases. And they cluster in certain regions. Most are in sections **A1** or **A2** which attach to actin filaments in the cytoskeleton. A few are in between these two sections. There is some presence in section **C** as well, where two copies of the *FLNA* protein attach to make a pair. But nothing at all in section **B**. Nothing at all, except the variant in *Som* and *Mira* that we have just found; it sits smack dab in the middle of this completely uncharted desert.

To summarize, many patients with missense variants in *FLNA* have been studied. None has a variant in section **B** so far. If variants in this region could cause disease, then surely we would have spotted such a variant in some patient by now. But we haven't. So, maybe *D1970N* is a mere onlooker rather than the agent responsible for the tragic fates of *Som* and *Rom*.

Or, maybe it does hide some dark secrets that we can't yet fathom. Our ability to study the genomes of a large number of patients effortlessly and inexpensively is very recent. Therefore, our knowledge of the variant landscape is still very incomplete. For instance, until 2010, only four *FLNA* variants causing *MNS* were known. All four were tightly clustered in section **A2**, prompting us to rush to the conclusion that only variants in this section cause *MNS*. In 2010, another missense variant causing *MNS* was discovered in between sections **A1** and **A2**,[51] reminding us to be patient and not jump to conclusions yet.

So, we keep our minds open on *D1970N*. But the desert around it yields precious little to establish its culpability. Yet, remember, there is something interesting about this desert section **B**. This is where *FLNA* interacts with many other proteins. Not just a handful, but several tens of proteins. The objective of this interaction is to convert mechanical cues to chemical cues.[46] This happens when mechanical shear forces acting on the cell force a larger angle between the two actin filaments, as in this picture.

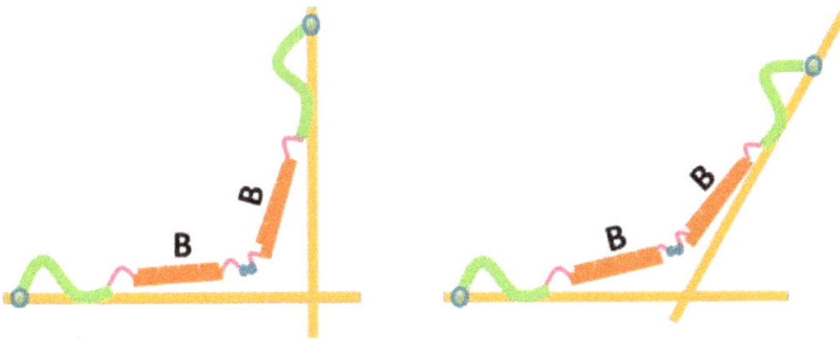

Deformation of cytoskeleton changes the angle between the two actin filaments, forcing FLNA to stretch.

As the cytoskeleton is sheared and two actin filaments are pulled apart, the two *FLNA* proteins attached to these filaments are stretched. Section **B** in these proteins is designed to absorb this stretch without snapping. It is composed of six units numbered 16–21. Usually, these units stack together in pairs, so the whole section is packaged very compactly. When the two actin filaments are pulled apart, these pairs respond by unraveling, as this picture illustrates for the 18–19 pair.

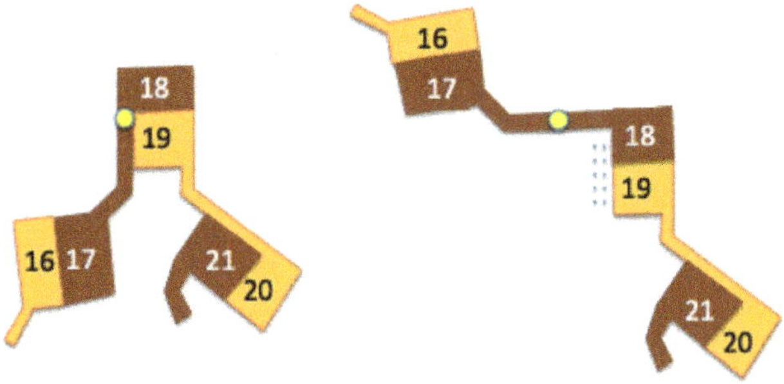

Units 16–21 in section **B** *on the left, and units 18 and 19 unraveled on the right. The circle indicates where our variant lies.*

Other pairs can open up similarly to yield even greater stretch. The important bit though: this unraveling causes parts of section **B** that were hitherto concealed to now be exposed to tens of other proteins. For instance, the dots in the picture on the right indicate parts of unit 19 that were hitherto covered by a tape-like extension of 18. These are now exposed and ready to react with many other proteins, leading to several chemical cues that play a key role in organ development. Thus, while **B** is a desert as far as variants reported in scientific literature go, it is eminently fertile when it comes to being in the scene of action.

Our variant of interest, *D1970N*, appears at the edge of the above tape-like extension that initially covers parts of unit 19. Could changing the *D* to *N* affect the opening and closing of this tape? Could it make *FLNA* more or less sluggish when reacting to mechanical forces on the cytoskeleton? Could it affect the interaction of section **B** with the tens of proteins it normally interacts with? Does the interaction with some important protein go haywire causing a key chemical cue to misfire? We have so many questions. Unfortunately, there are no ready answers.

We attempt to make some educated guesses by turning to a catalog of variants obtained from several tens of thousands of healthy individuals. We had checked this catalog earlier and verified that *D1970N* was absent from it. This time around, we train our eye on other *FLNA* variants in this catalog, hoping that they will whisper some secrets about their cousin *D1970N*. We look specifically at missense variants. There are 556 such variants in females and 291 in males. This difference is not surprising, for females have twice the number of X chromosomes as males. We study how these variants are spread across the span of the *FLNA* recipe.

The *FLNA* recipe has 2647 amino acids. We divide this span into 100 segments, each with about 26 amino acids. If variants in healthy individuals were spread evenly across this span, we would expect each segment to have 5.5 missense variants in females, and 2.9 in males. We check the actual numbers for the segment comprising amino acids 1961–1986, which contains *D1970N*. Five for females—that is close to par. For males? Just one, which, rather suggestively, is below par.

These are small numbers, but they do make a suggestion: variants here might have problematic consequences in males, so they rarely appear in healthy males. Their impact on females might be less problematic, thus explaining why *Mira* remains unaffected while *Som* was subjected to a tragic fate. That is a vote in favor of *D1970N*.

A look at the chemical compositions of *D* and *N* in this next picture suggests a prompt counter-vote though.

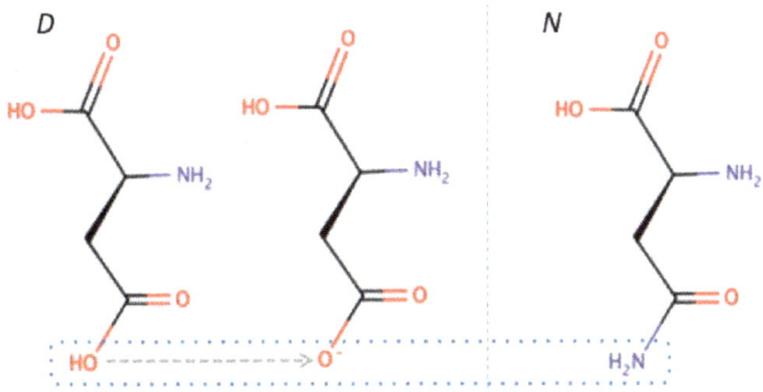

Amino acid D on the left, and N, on the right.

The two amino acids look almost identical, which means replacing one by the other ought to make little difference. There is a small difference though, right at the bottom—OH in *D* is replaced by NH_2 in *N*. Is this difference consequential?

It might be, in at least one respect. In the watery environment within a cell, *D* tends to lose a H^+ ion and thus becomes negatively charged as shown in the picture above. *N*, on the other hand, does not lose or gain any ions, and therefore remains uncharged. With this -ve charge, *D* can form stronger bonds with other positively charged amino acids than *N* can. *FLNA*'s ability to interact with other proteins depends on its ability to form bonds; possibly, *D1970N* weakens some of these bonds. There are indeed examples where an *D* to *N* change causes abnormality.[54] Another vote for *D1970N*'s potential culpability.

We do have a few votes now, but our case is far from strong. Let us investigate further and see if we can gather some more evidence that incriminates *D1970N*.

A Peek far into the Past

D1970N has never been seen previously in any other individual. Variants have been cataloged for tens of thousands of individuals; none of these has *D1970N*. Hopefully, many more people will have their genomes sequenced and their variants cataloged in the near future. If *D1970N* appears in a healthy male in this process, it would no longer remain a suspect, and we can move on. But this process will take time; what can we do in the interim? There is a trick. A clever one at that. It does not substitute for the inability to look at the genomes of all individuals alive today; however, in an indirect way, it actually allows us to look well beyond.

Each time a genome passes from parents to children, it changes a bit. These changes accumulate over generations. So, genomes of individuals appear more and more different as time goes by. By comparing the genomes of several individuals alive today, we can identify the extent of difference. Then, applying the rate at which differences appear on average from generation to generation, we can calculate how long ago our common ancestors lived. This calculation yields an estimate of about 200,000 years. We, the seven billion individuals on earth today, all appear to have descended from a relatively small group of people who lived 200,000 years ago. But aren't we digressing?

Hold on for a bit. We can go further than humans here and do this exercise with human genomes and chimpanzee genomes together. We sequence the genomes of a few chimpanzees, compare these with our various human genomes, identify the differences, and again work backwards to calculate when our common ancestors lived. Of course, this should be earlier than 200,000 years ago. A lot earlier, as it turns out—about 6 million years ago! By studying humans alone, we get a glimpse into only 200,000 years of evolutionary history. By studying chimpanzee genomes, we can get a glimpse into 6 million years of evolution.

Why just chimpanzees? We can cast a wider net, by doing the same exercise with elephants. This yields a common ancestor 100 million years ago. Of course, humans, chimpanzees and elephants are all mammals. We can go beyond mammals and consider the duck-billed platypus: half mammal and half egg-laying reptile—*metroneme* is the term for such creatures. Our analysis yields a common ancestor 200 million years ago. Snakes, which are clearly reptiles: 300 million years ago. Amphibious frogs, which live on both land and in water: 400 million years ago. And fish, which live fully in water: 450 million years ago.

Over these 450 million years, genomes have been passed down from parents to children, with each such step introducing a few changes. Darwin's famous theory of evolution suggests that these changes occur randomly in the genome, not in some calculated way and not with any specific intention. Yet, some of them stay and others are weeded out over time. Genomic characters that improve the fitness of a species tend to stay—these are *selected for*. Characters that compromise the fitness of a species tend to get weeded out—these are *selected against*. Characters that do not impact fitness at all could go any which way, depending upon where they lie in the genome. Fitness here means the ability to survive, find mates, and have offspring. Which brings us back to our point: over 450 million years of evolution, has the amino acid *D* at position 1970 in the *FLNA* recipe been selected for? Is this character important for survival?

Firstly, believe it or not, fishes, frogs, reptiles, metronemes and mammals all have an *FLNA* gene. What's more, the recipe encoded in this gene is strikingly similar in all of these very diverse life forms. Isn't that surprising? Well, all of these organisms have a challenge in common: they all have to form skeletons, hearts, brains and blood vessels starting from a single cell. This organ formation needs cells to change shape and move. *FLNA* is indeed a key player in this process. No wonder its recipe is strikingly similar across all these life forms.

The picture below shows these *FLNA* recipe sequences from various life forms stacked using a computer algorithm. This stacking of recipes allows us to compare characters across life forms, and thence answer our question: is amino acid D at position 1970 important for survival? Since these recipes are long, only a few amino acids around position 1970 are shown in this picture.

Mouse	RV-TGDDSMRMSHLKVGSAADIPINISETDLSLLTATVVPPS
Human	RV-TGDDSMRMSHLKVGSAADIPINISETDLSLLTATVVPPS
Frog	KI-TGDDSMRMSQLKVGSAADIPLNIVETDLSQLTATVTSPS
Zebrafish	KI-TGDDSMRMSHLKVGSAADIPLDIGELDLSQLTATLTTPS
Cow	RV-TGDDSMRMSHLKVGSAADIPINISETDLSLLTATVVPPS
Horse	RV-TGDDSMRMSHLKVGSAADIPINVSETDLSLLTATVVPPS
Rhesus Macaque	RV-TGDDSMRMSHLKVGSAADIPINISETDLSLLTATVVPPS
Duckbilled Platypus	KI-TGDDSIRMSHLKVGSAADIPLNITETDISQLTATVIPPS
Cat	RV-TGDDSMRMSHLKVGSAADIPINISETDLSLLTATVVPPS
Rabbit	RVTAGDDSMRMSHLKVGSAADIPINISETDLSLLTATVVPPS
Guniea Pig	RV-TGDDSMRMSHLKVGSAADIPINISETDLSLLTATVVPPS
Dog	RV-TGDDSMRMSHLKVGSAADIPINISETDLSLLTATVVPPS
Orangutan	RV-TGDDSMRMSHLKVGSAADIPINISETDLSLLTATVVPPS
Ferret	RV-TGDDSMRMSHLKVGSAADIPINISETDLSLLTATVVPPS
Bushbaby	RV-TGDDSMRMSHLKVGSAADIPINISETDLSLLTATVVPPS
Elephant	RSQVCDDSMRMSHLKVGSAADIPINISETDLSLLTATVVPPS
King Cobra	KI-TGDDTMRMSHLKVGSSADIPLNITETDLSQLTATVIPPS
Bat	RV-TGDDSMRMSHLKVGSAADIPINISETDLSLLTATVVPPS
Sheep	RV-TGDDSMRMSHLKVGSAADIPINISETDLSLLTATVVPPS

A comparison of FLNA recipes from various life forms around amino acid 1970, which is highlighted. Columns with dark text have the exact same amino acids in all organisms; light columns have at least two distinct amino acids.

Note the care needed in stacking these sequences. For instance, elephants and rabbits have an extra character in the third column— Q and T, respectively. Since none of the other organisms has this extra character, we must introduce an extra "-" in the sequences for these other organisms, otherwise elephants and rabbit recipes will be staggered relative to the others. With that done, our answer stares us in the face—amino acid D is common to all the organisms shown in this picture. Fishes, frogs, reptiles, metronemes, mammals—they all have a D at 1970. This D has been *conserved* for 450 million years of evolution!

Of course, by fishes, we mean just a few individuals from a few species whose genomes have been studied. Genomes have certainly not been studied for tens or hundreds of thousands of individual fishes. So, our fish *FLNA* recipe is just a randomly drawn representative; it is possible that other individuals have other amino acids at 1970. Nevertheless, if randomly drawn individuals from so many different life forms all show a D at 1970, it does suggest that the character D is an important one—if nature did experiment with a character other than D at this position, this new character probably caused a reduction in fitness and was therefore selected against and stamped out from its memory.

We cannot be entirely certain though that amino acid D at position 1970 is important for survival. It is possible that nature just hasn't had a chance in the last 450 million years to experiment with a different character at this position. So, let us dig a bit further into the past to give evolution a greater chance. It so happens that *FLNA* is a member of a family of related genes called *Filamins.* There are three genes in this family, *FLNA*, *FLNB* and *FLNC.* The recipes in all three are similar to each other, not 100% but 64%. The guess is that all three genes evolved from a common ancestor gene[55] that lived more than 450 million years ago.

How might this have happened? Remember that genomes are modified as they pass from parents to children. We saw one such modification, called crossing-over, in the very first story on my color blindness. If you remember, lopsided cross-over breakpoints in that story caused three genes to become four. Similar events could cause sections of the genome to be replicated, thus creating multiple gene copies from a single ancestor gene. This is possibly how a single ancestor gene lead to the formation of three distinct, but similar genes, *FLNA*, *FLNB* and *FLNC.*

Once formed, these three genes would have evolved independently.

Each would have experienced random changes, some of which would then have been selected for and some against. Over time, differences would have appeared between these genes. Again, based on these differences, we can calculate when the ancestor gene split into these multiple genes. The answer turns out to be about 500–550 million years ago.[55] By comparing *FLNA*, *FLNB* and *FLNC* amino acid sequences, we can push our window of observation into the past from 450 million years to 500–550 million years. This picture shows the comparison, focused again around position 1970.

```
FLNA  TGDDSMRMSHLKVGSAADIPINISETDLSLLTATVVP
FLNB  T-DDSRRCSQVKLGSAADFLLDISETDLSSLTASIKA
FLNC  TGDDSMRTSOLNVGTSTDVSLKITESDLSOLTASIRA
```

A comparison of FLNA, FLNB and FLNC around amino acid 1970. Columns with dark text have the exact same amino acids in all three genes; columns with light text have at least two distinct amino acids.

Amino acid *D* at 1970 is still conserved over 500–550 million years of evolution! Additionally, every one of the tens of thousands of humans whose variants have been cataloged has a *D* at this position, not just in *FLNA*, but in all three genes. Yet another vote for the culpability of *D1970N*. In an event that would certainly qualify as the rarest of the rare, *Mira*'s and *Som*'s genomes have an *N* instead.

Taking Stock

It's time to take stock and bring all the evidence we have against *D1970N* in *FLNA* to note.

The facial characteristics of *Som* and *Rom* were our first clue. Prominent eyes almost bulging from their sockets, full cheeks, and a small chin—these are all typical of Melnick-Needles syndrome or *MNS*. Both *Som* and *Rom* died suddenly of respiratory distress—also a common finding for patients with *FLNA* variants, including *MNS*.

Both had pulmonary hypertension—while not a common occurrence, this too has been reported in *MNS* patients. But *MNS* is typically found in females; males from affected mothers have severe skeletal malformations, and die before birth, or shortly thereafter. This was a bit puzzling at first sight, until we found reports in scientific literature of males from unaffected mothers with less severe manifestations—a description that would fit *Som* and *Rom*.

But there was a catch. Skeletal abnormalities are typical in *MNS*—bowed bones, a small rib cage, and ribbon-like ribs. Sometimes these are subtle, visible only under X-rays. *Som* and *Rom* certainly did not have any obvious skeletal abnormalities. And they were not investigated for more subtle abnormalities because their visible symptoms did not suggest the need for a thorough skeletal investigation. Of course, their doctors had no inkling that an *FLNA* variant might be at play. Sadly, by the time our genomic investigation started, *Som* and *Rom* had passed away.

This lingering element of doubt made us look very carefully at the *FLNA* variant *D1970N*. We looked at tens of thousands of humans whose genomic variants have been cataloged. We also traveled 500–550 million years back in time looking at the genomes of several other organisms. Yet, we were unable to spot anything but a *D* at location 1970. We also examined other variants in the vicinity of 1970 in healthy males and found lesser variation than would be expected. Clearly, something was resisting change in and around position 1970. All of these facts together suggested that *D1970N* could have problematic consequences in males, of the type we saw in *Som* and *Rom*.

Yet, we have no idea how these problematic consequences might come about. Whether *D1970N* affects the opening and closing of the tape-like extension of unit 18 as it unravels when stretched, or whether it affects the interaction of several tens of proteins with *FLNA* leading to the misfiring of some chemical cues of importance in organ development, we do not know.

We do know that *FLNA* lies on the X chromosome, of which males like *Som* and *Rom* have just one version. There is a 25% chance that both boys inherited *D1970N* from their mother *Mira*, who has this variant in one of her two gene versions. Had this gene been on one of the other 22 chromosomes, both boys would have needed to inherit problematic variants from both parents—a chance of only 6.25%, much smaller than 25%. So, a gene on the X chromosome is indeed the most likely candidate from this perspective.

If only we could access *Rom*'s genome, we would have checked it for *D1970N*. Its presence would have added another vote, while its absence would have ruled it out as a suspect. However, *Rom* had passed away long before, and his genome was out of reach. The closest substitute was to check other males in the extended family instead. Presence in a healthy male would rule out *D1970* as a suspect. Unfortunately, *Mira*'s father too had passed away much earlier. However, his genome lives on indirectly in his children *Mira* and her sister *Sara*. So, we checked *Sara*. She had no signs of *D1970N*—both her *FLNA* recipe versions have a *D*. One of these two versions was inherited from her father. So, *Mira*'s father did not have *D1970N* either.

Therefore, *Mira* had either inherited *D1970N* from her mother, or had acquired it *de-novo*—via a copying error that arose when she was conceived. Her mother too was long deceased, as were most of her immediate male relatives, but for one exception: *Mira*'s husband *Sai*. Remember, this was a consanguineous couple—*Sai* was *Mira*'s mother's brother, or her uncle. He, of course, did not have the variant. In summary, the search for a healthy male with *D1970N* remains elusive, keeping the needle of suspicion pointing at this variant.

There is one key symptom in *Som* which we haven't addressed at all: systemic hypertension, or high blood pressure in the arteries that carry oxygenated blood from the heart to the body. While patients with *FLNA* variants do exhibit pulmonary hypertension—high blood pressure in the artery carrying blood from the heart to the lungs—none has been reported with systemic hypertension yet. Tests had ruled out the obvious heart, kidney or hormonal causes of hypertension in *Som* and *Rom*. The presence of both systemic and pulmonary hypertension in these boys then suggests that the cause might lie in their blood vessels.

Blood vessel cross-sections are composed of many layers. The innermost layer comprises *endothelial cells* that release *nitric oxide*. The next layer comprises smooth muscle cells which react to nitric oxide by contracting or dilating. This contraction and dilation controls the pressure exerted by blood against the vessel walls. Elastic arteries that dilate and contract easily keep this pressure under check. Stiffer arteries that cannot dilate and contract so easily may face higher amounts of pressure. Other structural artefacts like abnormally narrow arteries could also contribute to high blood pressure. Is it implausible that an *FLNA* variant can produce any of these abnormalities?

Not really. By its usual modulation of cell shape and movement, *FLNA* plays a key role in the formation of these various layers, the endothelial layer in particular. We know this because mice bred in the complete absence of the *FLNA* gene have a defective endothelial cell arrangement that causes widespread hemorrhaging.[56] Possibly, a pointed *FLNA* variant like *D1970N* may yield a less defective structure—blood vessels may not hemorrhage but their ability to contract and dilate as usual may be impacted, leading to hypertension.

That is our dossier on *FLNA*'s *D1970N*. The evidence in this dossier is strong enough that multiple experts when presented this evidence agree that it appears highly suspicious, at the very least. But, this isn't conclusive scientific proof. That requires experiments where *D1970N* is introduced into a living system and its consequences studied—a process that could take months, if not years, not to mention sizable funds. And time is what we don't have...

Will *FLNA* strike again?

Before long, *Mira* and *Sai* have decided to move forward, putting their previous tragedies behind them. *Mira* is now pregnant with their next child. Of course, the last thing the couple wants is a repeat of the traumatic experience they underwent with *Som* and *Rom*. Would they have to wait until their child is a year old or more before they know what lies in store? Or could we help them expedite this agonizing wait with each moment spent in fear of impending tragedy.

For several years, medical science has had the ability to extract the baby's genome prenatally, while it is still inside the mother's womb. Typically, a person's genome can be obtained from his/her saliva or blood. In fact, even a sample of hair or fingernail will do. But how does one get any of these samples from a tiny foetus deep inside its mother. There is indeed a way.

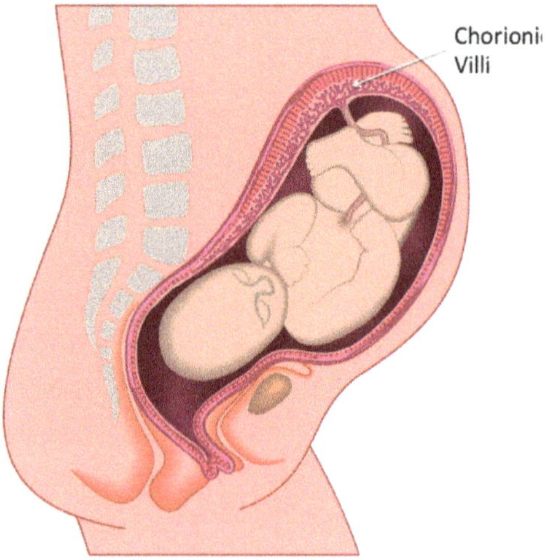

Chorionic villi in the placenta.

The umbilical cord from the baby is attached to the mother through an organ called the placenta. It is through the placenta that the baby gets its nutrients and disposes of waste. A part of the placenta called the *chorion* comes from the baby and carries its genome. The other part of the placenta called the *decidua* comes from the mother. To pose the least risk to the baby, a little tissue sample is taken from tiny protuberances on the chorion called *chorionic villi*, shown above.

This is done by a needle inserted into the abdomen under the watchful guidance of an ultrasound probe. The sampling procedure—aptly called *CVS* or chorionic villus sampling—can be performed when the baby is just 10 weeks old. It must be done carefully though, otherwise the sample extracted could contain a mixture of the mother's and the baby's cells, and genomic analysis could go astray.

CVS does pose a small risk to the pregnancy. So, it is performed only in high-risk pregnancies—when there is a family history of a genetic disorder, or when ultrasound screens indicate something is amiss. For instance, an ultrasound may sometimes indicate *nuchal translucency*—the baby has too much fluid collecting around the neck. This is suggestive of *Down's Syndrome*, which is caused by the presence of three versions of chromosome 21 in the baby instead of the usual two versions. If nuchal transparency is seen in an ultrasound at the end of 12 weeks, a CVS sample is taken and a genetic test is performed to count the number of versions of chromosome 21. An abnormal result provides the family an option to terminate the pregnancy early, or equally well, set their expectations in accordance with a differently-abled child.

In recent times, a CVS sample is being replaced by an even less riskier method: mother's blood. Mother's blood has free floating genomic fragments, 5–10% of which come from the baby she carries, and the remainder from the mother herself. Unfortunately, these fragments are all mixed up, so we don't know which fragment came from whom. Regardless, there are clever ways to determine whether the baby has three chromosome 21 versions or two, just by sequencing all these fragments collectively. However, the use of mother's blood is not quite as well-developed when it comes to identifying a pointed missense variant, like *D1970N*, in the baby.

Getting back to our story, *Mira* is in the 12th week of her pregnancy now. Clearly, there is a strong family history of disease to justify a CVS. Added to that, an ultrasound scan in the 12th week shows nuchal translucency and a depressed nasal bridge, both of which are indicative of Down's syndrome. A CVS is promptly performed. The number of copies of chromosome 21 is then checked. There are exactly two copies in the baby, so all is well so far. The question that hangs heavy on everyone's mind: does this baby also have *D1970N* and will it encounter the same fate as its deceased brothers?

What if the baby turns out to be male with the *D1970N* variant in his sole copy of *FLNA*? The family must undergo genetic counseling and be made aware of possible future consequences—possibly, or even probably, though we cannot be fully certain, this baby will follow the same trajectory as its ill-fated brothers. Like most families who can only fathom so many shades of gray, this one will turn to their clinician for her recommendation. The clinician will have a difficult decision to make. On the one hand, could she let the pregnancy continue and subject this family to the threat of yet another tragic event? They were looking to her for help, and she finally had a tool at hand to provide that help. However, this tool came without the assurance of certainty. Could she ignore this tool and see them plunge into sorrow again? On the other hand, could she recommend early termination of pregnancy without complete conviction?

For the clinician, this is her first experience with the use of large-scale genome sequencing to diagnose what appears to be a mysterious disease. It is also early days for the entire medical community—barring some thought leaders and researchers, few in regular medical practice have used it for making these delicate and difficult decisions. In all likelihood, this is the first case of its kind in India. How will a drastic decision such as early termination of pregnancy be viewed in this light? Since the culpability of *D1970N* has not been proven conclusively, what happens in the off-chance that it really isn't the culprit? What if an autopsy on the aborted foetus shows that all is well suggesting that the foetus was aborted for no reason? There are also strict laws in India against revealing the gender of the baby before it is born. Given *FLNA* lies on the gender-determining X chromosome, how can the clinician take this decision without revealing the gender? All tough questions, as the critical moment approaches fast.

Wrapping Up

For several diseases, skilled geneticists can actually pinpoint the underlying defective gene without so much as a peek into the genome. They do this by simply matching symptoms to disease descriptions. This needs a very sharp, trained eye. However, from time to time, a new combination of symptoms presents itself, challenging even the sharpest and best trained of these eyes. A genetic diagnosis remains elusive in such cases making families such as *Mira*'s struggle with the tragic prospect of giving birth and raising a child that they might soon lose, yet again. Against this backdrop, the ability to look deep into the genome for the root cause provides a tool that empowers families with a choice very early in their pregnancy. For *Mira* and her family, the identification of *D1970N* provides such a tool.

Several different lines of evidence had to be brought together to identify *D1970N* as the likely candidate, including some clever thought experiments that took us 500 million years back in time. However, unlike in previous stories where we could pin down genes conclusively, we do not have conclusive proof for *D1970N*. Indeed, conclusive proof for such *variants of unknown significance* that have never been observed in anyone else, or for that matter in any other organism for the last 500 million years, requires expensive and time-consuming scientific experimentation. On occasion, researchers perform these experiments proactively, bringing some very difficult cases to a quick conclusion, as we will see in the next story. In general though, this takes months if not years, and the results could go either way. But *Mira*'s pregnancy cannot wait—time is running out and her clinician must make the best decision possible, taking all evidence, and all ethical and legal concerns, into account. What a tough decision to have at hand.

The pregnancy has reached its 15th week. The ultrasound still shows a depressed nasal bridge, indicating potential lack of normalcy. A decision has to be made soon. Genetic results for *D1970N* arrive in the 16th week. They show that the baby has two versions of *FLNA*, one with a *D* and another with an *N*, just like its mother *Mira*. Further, just like its mother, it has lost the start indicator from only one of its two *FBLN5* versions. By the 17th week, signs of a depressed nasal bridge also clear up on the ultrasound. There is every indication now that this will be a healthy baby. So, the pregnancy continues. A little girl is born uneventfully several months later. The parents are happy to note that her eyes aren't bulging out conspicuously as they did in her ill-fated brothers. When last checked a year later, she appears normal on all counts. And blissfully unaware of the history she has created, as the first child in India whose birth was supported by an army of biologists and computer scientists scouring every inch of her family's genome sequence in a bid to ensure that she lives a healthy life.

Story 5

Not Quite a Mirror Image

This is rather an unusual story of twins, *Raj* and *Taj*. To set the context for their condition, take a moment to review how the various organs are placed in our bodies. The liver is typically on our right and the stomach on our left. *Raj*, on the other hand, was born with his liver on his left and his stomach on his right, as in this picture.

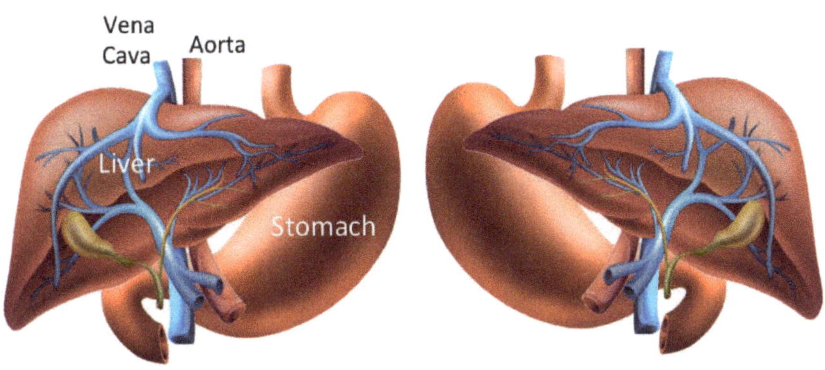

The normal anatomy, on the left, and Raj's anatomy, on the right.

Further, note that the *aorta* typically runs down along the left of the backbone, and the *vena cava* runs up along its right. In case you are wondering, the aorta is the main artery that takes blood from the heart to the body, and the vena cava is the main vein that takes blood back from the body to the heart. Oddly enough, *Raj*'s aorta ran down along the right of his backbone, and his vena cava ran up along its left. *Raj*'s anatomy thus appeared to be a rather surprising mirror image of its usual self.

Though odd, such mirror flips in anatomy may not always have stark consequences. For instance, some organs like the lungs and the kidneys come in pairs that are almost symmetric anyway. Others like the urinary bladder are solitary, yet symmetric. Had *Raj*'s heart also been flipped, the various organs would have remained consistent relative to each other. Unfortunately, this was not the case. *Raj*'s heart wasn't flipped around and this caused inconsistencies. In particular, the wiring of blood vessels into his heart had gone completely awry. Such inconsistency in organ placement has a name—*heterotaxy.*

To appreciate the impact of this heterotaxy, recall how blood flows through the heart. The deoxygenated blood carried by the vena cava flows through the right side of the heart to the lungs, and the oxygenated blood from the lungs flows through the left side of the heart into the aorta and thence to the rest of the body, as shown below.

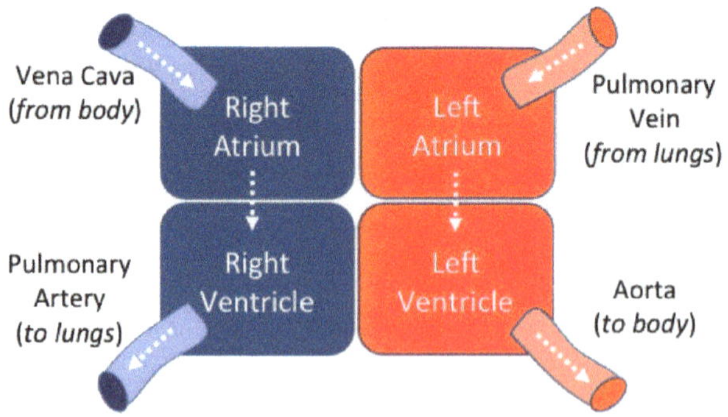

The normal blood flow through the heart. Deoxygenated blood flows through the right, and oxygenated blood through the left.

The wiring in *Raj*'s heart was different. His blood vessels were flipped, so deoxygenated blood flowed through the left to the lungs, and oxygenated blood flowed through the right to the rest of the body, as in the picture below.

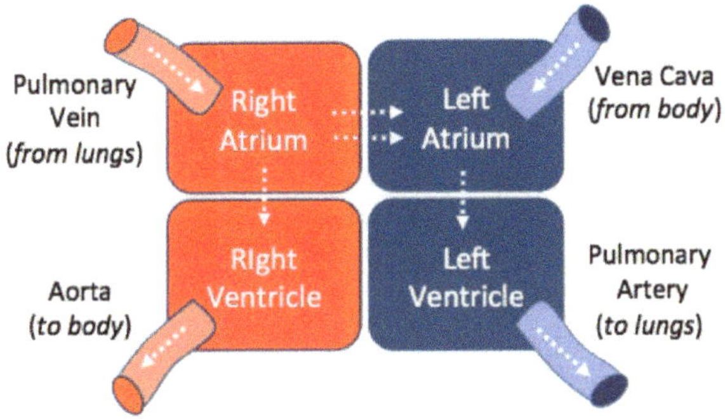

Abnormal wiring of blood vessels into the heart in Raj.

As you can imagine, it takes more effort to pump blood to the whole body than it does to pump to the lungs alone. Accordingly, the walls of the left ventricle are thicker and more muscular. In *Raj*, the right ventricle had taken on the onus of pumping blood to the whole body. Since its walls are not as muscularly designed, this abnormal wiring would subject it to excessive stress. Conversely, the more powerful left ventricle would now put excessive pressure on the delicate blood vessels in the lungs.

Unfortunately, that wasn't all. Heterotaxy comes with additional complications, and *Raj* was no exception. As the picture above shows, the wall separating the left and the right atria had a hole through which blood leaked from the right atrium to the left. This meant that some of the oxygenated blood from the lungs entering the right atrium found its way back to the lung in a short circuit via the left atrium, the left ventricle, and the pulmonary artery. Only a portion of this oxygenated blood actually flowed to the rest of the body, depriving it of much-needed oxygen and causing parts of it to turn blue.

Raj's brother *Taj* was six now. He too had been born with similar, though not identical, alterations in anatomy. In both children, surgery on the heart to fix these defects was necessary for survival. Both parents and other siblings of *Raj* and *Taj* were normal and conformed to the standard architectural plan. How then did *Raj* and *Taj* acquire this unusual architecture?

Where Do We Look?

Presumably, the answer lies in their genomes. Our challenge is to identify the offending genomic character(s) by sequencing the genomes of *Raj*, *Taj* and their parents. In previous stories, we extracted and sequenced the exons of all 20,000 genes from our subjects. We chose to sequence only the exons while ignoring introns and other vast intergenic regions that lie between genes because gene recipes are carried only by the exons. For *Raj* and his family, the whole genome was sequenced—introns, intergenic regions and all. And this difference will be instrumental in our quest, as we will see soon.

Of course, sequencing the entire genome comes at an added cost. This is often unnecessary, because most problematic variants causing disease are present in exons. However, in some situations, sequencing the entire genome can be a life saver. Among other things, it is actually quicker, for isolating the exons from the rest of the genome takes a few days longer in the laboratory than simply going ahead and sequencing the entire genome. For instance, Stephen Kingsmore and his team at Children's Mercy Hospital, Kansas City, have pioneered the practice of rapid sequencing for diagnosing children with genetic defects[57]—they sequence the entire genome in 24 hours, and deliver a diagnosis after analyzing the resulting variants in another 24 hours! *Raj* and his family comprised one such case. A particularly difficult case though, as we shall see.

Sequencing the entire genome also generates an enormous amount of data—100 gigabytes for each subject. Yet, data crunching on a computer can be performed in a few hours with clever algorithms and systems, and yields about four to five million genomic variants per person, as opposed to tens of thousands in the previous stories. Remember, variants are positions where the subject's genome differs from the reference sequence—the genome sequence of a few healthy individuals. Which of these four to five million variants is the culprit here?

We start by narrowing down our list of millions of variants to genes of interest. Our first shot is at genes well known to cause shades of mirror image anatomy. Scouring through scientific literature yields a handful of such genes. As always, we look for variants that have a significant impact on the gene recipe, ignoring others which leave the recipe untouched. Of these, we look only for rare variants which do not appear in many people, because common variants are unlikely to cause a rare condition such as heterotaxy. Our experience investigating *Som* and *Rom*'s early deaths in the last story then reminds us of the inheritance scenarios we need to consider.

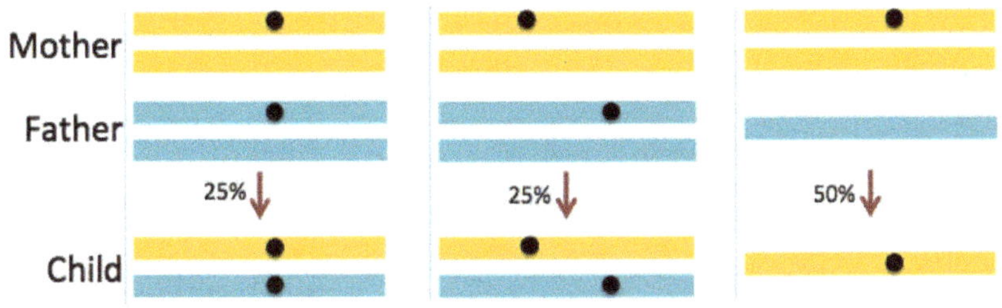

Various inheritance scenarios; the last scenario involves the X chromosome.

Accordingly, we retain only variants that are found in both *Raj* and *Taj* and that satisfy one of these inheritance scenarios. Of course, *Raj* and *Taj* are both boys, making the X chromosome scenario the one with greatest likelihood. Is the culprit in this case also a gene on the X chromosome? We let our computer loose on our long list of variants to get the answer. The computer takes a moment—we hold our breath. And then it returns a blank. None of the genes known to cause shades of mirror image anatomy seem to apply to our case.

Our next shot is at genes which are known to cause some abnormality in body development that is serious enough to manifest at birth or in early childhood. This list comprises a few thousand genes like *FLNA* from the last story. Again we look for variants in these genes as we did for the handful of genes above. And yet another blank. Here we have a family waiting for a diagnosis, and the state of the art offers us no help at all. There is little alternative but to foray bravely into virgin territory now.

We move on to the remaining 17,000 or so genes, none of which are known to cause even the remotest of architectural changes, let alone changes of the type seen in *Raj*. We instruct our computer to wade through the millions of variants in *Raj* and *Taj* and look for variants with the properties described above. The computer takes a little longer this time. Yet another blank? Thankfully, not this time. One gene appears on the screen—a gene called *BCL9L*. Could this be the one instrumental in placing *Raj*'s liver on the left, and his stomach and his spleen on the right?

Left-Right Symmetry

Before we investigate *BCL9L* further, a fundamental question poses itself—why is the stomach on the left in almost every one of us? Note that nature is not picking randomly between right and left; had it done so, several of us would have our stomach on the left and several others on the right. Nature somehow knows left from right so definitively that it gets the answer correct almost every time. How does it do that?

Maybe we can take a step back, and ask an even more fundamental question: why are our bodies shaped the way they are, and not as perfectly symmetric round balls? This is because symmetry-breaking events abound in nature. Here is one example.

We begin life when a single sperm cell from our father fuses with an egg cell from our mother. After the sperm and egg cells combine, the distribution of various molecules inside the combined cell becomes asymmetric—a different mix of molecules appears at the point where the sperm combines with the egg, as compared to the point diametrically opposite. This asymmetry then compounds to yield pronounced differences between the two ends. Indeed, in frogs, the sperm entry point develops into the back and the opposite point develops into the belly,[58] as indicated in the picture below.

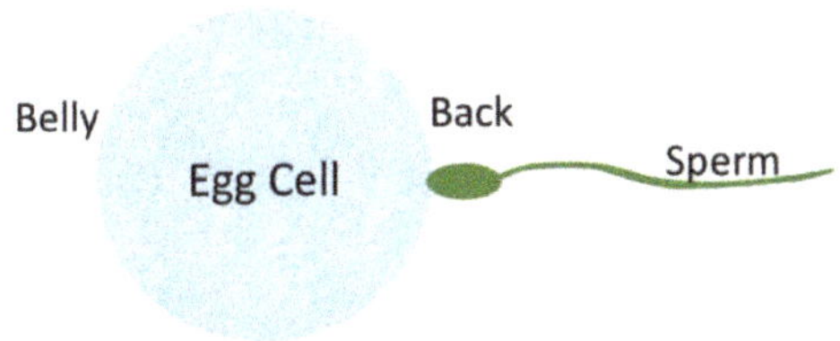

The sperm attachment point breaks back–belly symmetry in frogs.

In humans and mice, the sperm attachment point is less important and symmetries are broken via more complicated events. Nevertheless, symmetries do get broken. In fact, since we live in a three-dimensional world, we have three axes describing our bodies, and symmetry is broken along all three: the head–toe axis, the back–belly axis and the left–right axis.

If you think about it, the lack of symmetry along the first two axes is patent—the head looks very different from the toe, and the back looks very different from the belly. The left–right axis is subtle, though.

From the outside, our left looks much the same as our right, barring minute differences. The placement of internal organs is highly asymmetric, however. This asymmetry is the last to set in, after the back–belly and the head–toe symmetries have been broken, and therein lies a mystery.

We may not care if nature picks the sperm entry point as the back end and the opposite point as the belly end, or vice versa. Whichever option it picks, one end will become the back, the other the belly, and we'll look much the same if we orient ourselves with the belly to the front. Likewise, we may not care if nature interchanges the head and toe ends—we can always orient our bodies so the head comes out on top and the belly on the front. But once we orient ourselves so, nature has to be more careful with the left–right axis.

If nature mixes up left and right now, then the liver might go on the left and the stomach on the right. No amount of reorientation will get the liver to the right, and still keep the head on top and the belly to the front.

Surprisingly, nature rarely makes this mistake. It almost always picks left from right correctly, placing the liver on the right in the process. Hence the mystery: how does nature know left from right, once the back–belly and the head–toe symmetries have been broken? How, indeed?

Much of what we know about this question comes from studies in other animals, with mice being closest to humans.[59] Imagine a mouse embryo and suppose various symmetry-breaking events have blessed it with distinct head and toe ends, as well as distinct back and belly ends. This picture shows a rather rough outline of such an embryo, 7.5 days old.

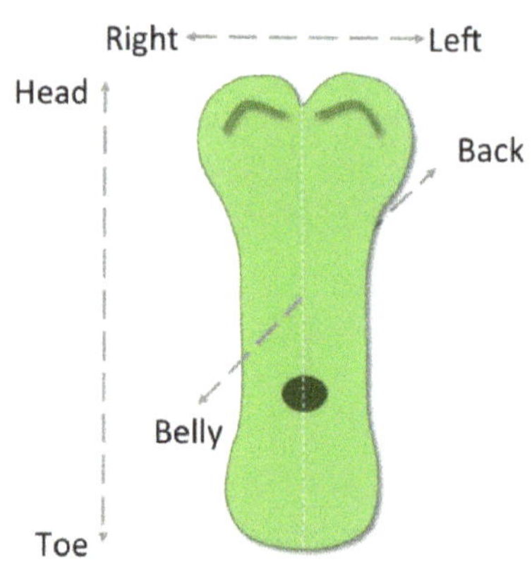

A very rough outline of a 7.5-day-old mouse embryo.

This embryo appears to be left–right symmetric as far as we

know, both in shape, and in the distribution of various molecules. However, this symmetry is short-lived, for a symmetry-breaking event occurs soon. As a consequence, certain molecules, like the one called *nodal*, become more preponderant on the left than on the right in an 8.5-day-old mouse embryo, as in this picture.

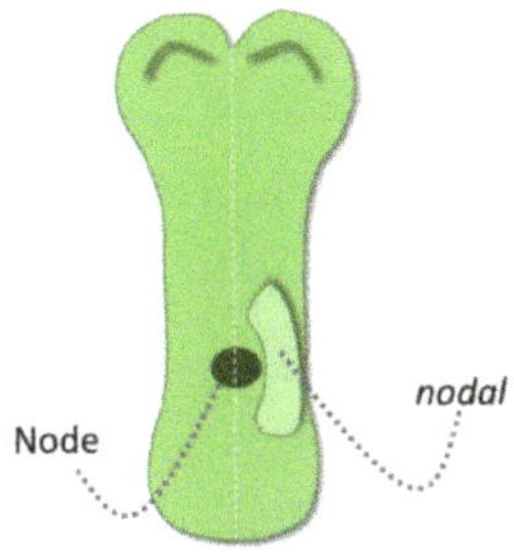

An 8.5-day-old mouse embryo. The nodal *protein preferentially aggregates to the left of the dark patch labeled* node, *which comprises cells that will shortly turn out to be important.*

Then onward, this asymmetry cascades, with *nodal* instigating several other proteins differentially on the left. Eventually, this causes some organs to form on the left and others on the right. The question then is: what is this symmetry-breaking event? How does this event know what is left and what is right? Does this information come from within the embryo itself, from the mother, or from the world more broadly?

The notion that it comes from the embryo itself was shown by studies which introduced specific variants into mice genomes.[60] These variants interfered with the left–right symmetry-breaking process. As a consequence, half the individuals in these studies had symmetry broken in the usual direction, so the liver appeared on the right, and the other half had symmetry broken in the opposite direction, so the liver appeared on the left.

Importantly, the direction in which symmetry was broken above was independent of which side the mother's liver was placed; roughly half the children from the same mother had their liver on one side, and the other half had it on the other side. So, clearly, the symmetry-breaking process appears to be controlled by the embryo itself and quite independent of the mother and the world at large.

This leads to the next question. How does the embryo know by itself what is left and what is right? A moment's pause at this point may allow this question to sink in before we get to the answer, which is fascinating, and rather clever.[59]

The scene of action here is a clump of cells bathed in a fluid and labeled as the *node* in the picture above. Each such cell carries a single *cilium*, a hair-like projection, at its belly end.

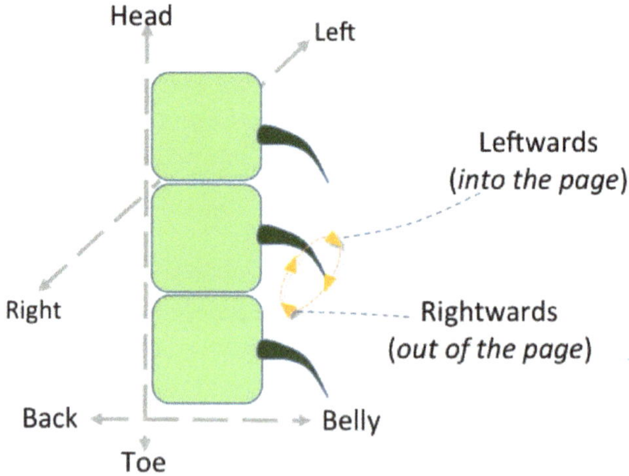

Cells of the node, in profile, and their cilia.

These hair-like cilia might appear small and unimportant at first glance. Deceptively so, for they play critical roles in the development of our various organs. Indeed, defective cilia have been established as the primary cause for at least 13 different diseases. In addition, they are suspected to be associated with 120 other diseases as well.[61]

A hallmark of these cilia is that they are equipped with molecular motors which make them move in clockwise circular beats. You will notice that the beats go into the page at the head-end and emerge out of the page at the toe-end. All the cilia share this same clockwise motion. Together, these circular beats push the surrounding fluid around, thus setting the stage for the embryo to break left–right symmetry.

Of course, so far, left–right symmetry has not been broken. Now, here's the magic. The collective circular motion of all the cilia pushes the fluid around these cells to the *left* (note: *specifically to the left*)—regardless of whether the embryo and mother are facing north, south, east, or west, and sleeping, standing, or walking. The surrounding fluid is always pushed toward the left. And there goes left–right symmetry.

But why does the circular motion of cilia push the fluid to the left, and not to the right? Isn't there a leftward stroke at the head end and a corresponding rightward stroke at the toe end? The former should push the fluid to the left and the latter should push it back to the right. Together, these opposing pushes should cancel out, yielding net zero flow of the fluid.

A small yet critical geometrical artefact prevents a complete cancelation—note that each cilium points not straight ahead but at a toe-ward angle. Watch its circular trajectory carefully now. The leftward stroke occurs farther away from the cell surface than the rightward stroke. Now the key point: surfaces offer friction, so fluid movement is freer farther from the cell surface than it is closer to the surface. This difference in friction causes the leftward stroke to move the fluid more easily than the rightward stroke. As a net effect, the fluid gets pushed to the left, thus breaking left–right symmetry.

Why should we believe this story? Because a specific genomic variant that cripples the motor that makes cilia move has been introduced in some mice. Lo and behold, such embryos break symmetry randomly, some in favor of left, others in favor of right, and yet others partly in both directions. The *nodal* protein mentioned above is found on both sides in the latter case. Accordingly, some of these embryos have normal organ placement, others have mirror-image placement, and yet others show heterotaxy with a mix of normal and mirror-image placement, as in our boys *Raj* and *Taj*.[60] Even more convincingly, when the fluid is moved by artificial means to one side in embryos with the above genomic variant, symmetry is broken consistently toward *that* side.[62]

So, most unexpectedly, fluid being pushed around by cilia breaks left–right symmetry. There is active debate on other events which might break symmetry even earlier.[63] Regardless, when genomic variants interfere with this process, the body plan is altered—either as a complete mirror image, or as heterotaxy. Such genomic variants are known in several genes. Unfortunately, *BCL9L* is not one of these. Could it really be the culprit here?

BCL9L: Is it the Culprit?

As far back as 1910, Thomas Hunt Morgan who studied fruit flies (flies that hover over uncovered fruit), noticed an unusual fly: one with white eyes instead of the usual brilliant red. His pioneering studies exploring the inheritance of these white eyes established a long tradition of such genetic studies in fruit flies. Since then, scientists have bred fruit flies with various odd traits—orange eyes, legs coming out of their heads, curly wings etc.—and traced these traits back to the causative gene in the genome.

One such trait is winglessness. Researchers observed some wingless flies in the 1970's and eventually traced the cause back to a gene which they named *wingless*, for obvious reasons. Further research on the *wingless* gene showed that it sets off a cascade of events in the developing embryo, now called *Wnt Signaling*. This cascade plays a crucial role in breaking symmetry along the head–toe axis as well as the back–belly axis.[64,58] In that process, it also plays a role in the formation of the cilia whose clockwise motion creates left–right asymmetry.[65]

Our suspect *BCL9L* appears somewhere in this cascade,[66] making us hopeful. But precious little is known about its exact role. Hoping for some clarifying insight, we turn toward the specific variants that

Raj and *Taj* have in this gene.

Both brothers have inherited one variant from their mother, and a different variant from their father. Each parent has one gene version that is normal, thus protecting them from the disease. In contrast, *Raj* and *Taj* have variants in both gene versions. Are these variants indeed problematic variants? Or are they innocent bystanders who just happened to be at the scene of the crime?

Unfortunately, both variants are missense variants, making this assessment difficult. One of these variants is *A185V*—amino acid *A*, short for *Alanine*, at position 185 in the recipe, is replaced by amino acid *V*, short for *Valine*. It appears to be quite rare, having been seen previously in only a few healthy individuals among several tens of thousands whose genomic variants are archived in various databases. Even these individuals have the variant in only one gene version; the other version appears to be in good shape, possibly explaining why they are healthy. This is a possible vote for this variant's culpability.

To collect more evidence, we attempt the time-travel thought experiment we used in the last story where we investigated the early death of *Som* and *Rom*. In the same manner, we compare the *BCL9L* gene recipe across several organisms to identify whether nature has experimented with alternative amino acids at position 185 over the last hundreds of millions of years. We find that most mammals indeed have an *A* at this position. But a marsupial, the gray short-tailed opossum, has a different amino acid, a *G*, short for *Glycine*. Fish also appear to be different in and around this position. So, nature appears to have experimented with *A* at position 185 and got away with it. This is not much of a vote for *A185V*'s culpability.

The other variant in *BCL9L* is *G701D*, a change from amino acid *Glycine* to *Aspartic Acid* at location 701. Again, birds, reptiles and fish don't have a *G* here. Neither do some mammals like dogs and cats. So, not much of a vote at all.

But *BCL9L* is our only suspect. As Sherlock Holmes might have said: *once the impossible has been eliminated, whatever remains, however improbable, must be the truth.* So, however weak the evidence might be, *BCL9L* remains our prime suspect; we just have to verify its culpability by experiment. To do this, we have to introduce these variants in a mouse or another animal, and observe their impact on their anatomy, however time-consuming that might be. So, we brace for the long haul.

Mice to the Rescue

While we were holding, and quite unbeknown to us, scientist Cecilia Lo's group at the University of Pittsburgh had embarked on an ambitious study to uncover more genes responsible for heart malformation of the type that *Raj* and *Taj* had.

The study aimed to create variants in mice genomes by treating mice with a chemical called *ethylnitrosourea*, or *ENU* for short. ENU introduces variants randomly in the genome, once every 1000 or so characters, in the sperm cells of male mice. Many of these variants are harmless. Some, however, might be capable of causing heart malformation. The hope was that a few of these would arise by sheer random chance and provide some insights into genes that play a role in the development of the heart.

Of course, even those variants capable of causing heart malformation would typically do so only if they were present in both gene versions. However, the random effects of ENU treatment are unlikely to introduce variants in both gene versions simultaneously. And that raises the following question. Starting with a mouse with a problematic variant in one gene version, how does one create mice with problematic variants in both gene versions?

For centuries, breeders of all kinds, including plant breeders and dog breeders, have had an answer to this challenge. Cecilia Lo's group followed a similar prescription.

The group mated an ENU-treated male mouse, which carried a particular variant in one gene version, with standard female mice. Half the resulting offspring inherited this variant from their father; these offspring now carried the variant in exactly one of their two gene versions. The females among these offspring were mated again with their father. Since some of these carried the variant in exactly one of their two gene versions, their further offspring had a quarter chance of inheriting variants in both gene versions. Sufficiently many matings then yielded several mice, each with the variant in both gene versions. The picture below illustrates this strategy.

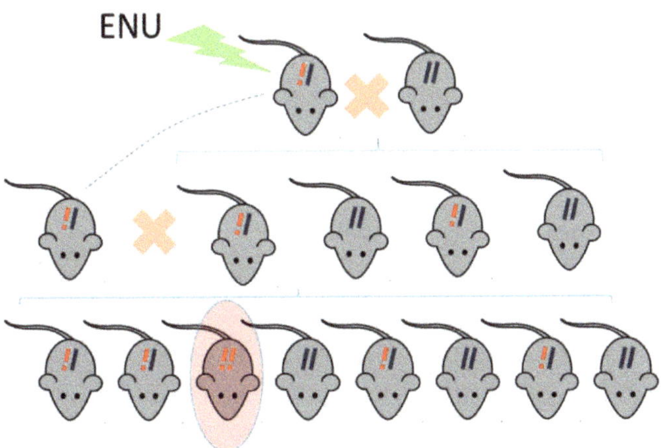

Crossing an ENU-treated male mouse with normal female mice, across two generations. Genome versions that carry a variant introduced by ENU are shown as broken stripes, while normal genome versions are shown as solid stripes.

Of course, being a random process, the chances of ENU creating a variant causing heart malformation are not that high in the first place. So, the above process was repeated with many many male mice. This was truly a mammoth exercise. Cecilia Lo's group scanned each resulting mouse for heart malformations—an impressive 14,000 mice in all! Finally, the genomes of any mice found with malformations were sequenced to identify the causative variant and gene.

This marathon effort yielded several new genes causing heart malformations. Not unexpectedly, many of these genes were involved in the formation and functioning of cilia, which if you remember were instrumental in breaking left–right symmetry. Disappointingly,

BCL9L was not on this list.

What Else is Possible?

Sherlock Holmes's words could still ring true: *once the impossible has been eliminated, whatever remains, however improbable, must be the truth.* *BCL9L* could well be the answer, just that random variant generation by ENU happened not to introduce the right variant in *BCL9L*, and therefore, the gene did not appear in the list above. But there is a catch in Holmes' claim. Unless we have considered *every possibility*, we cannot eliminate all that is impossible. Have we really considered every possibility?

We carefully single-step through our assessment of variants in *Raj*'s and *Taj*'s genomes again. We considered only rare variants that modify gene recipes, either by truncating them prematurely or by introducing one amino acid for another. Should we consider variants that are not rare, and found in many normal people? No, for these are unlikely to cause a rare condition such as heterotaxy. Should we consider other types of variants? There are indeed other variants that do not modify the recipe directly, yet influence it indirectly. Some of these will be central themes in the stories that follow. We do check for these types as well but find nothing of significance.

What else could we have missed? We insisted that our variants of interest be found in both *Raj* and *Taj*. That's reasonable, for both show the same symptoms. We also insisted on one of the recessive inheritance scenarios—the children must have variants in both gene versions, while the parents must have variants in just one gene version each. Did we miss something here?

Could one of the parents not have any variants at all? Is that at all a possibility?

How can that be? The children must have variants in both gene versions for a recessive disease to occur. One gene version is inherited from each parent, so each parent must carry a variant, shouldn't they?

There is, of course, the *de novo* possibility—variants need not always be inherited; they could be introduced due to copying error when the sperm or egg was created. However, that is a rare enough occurrence and very unlikely to have happened for both children. So, do we really need to look for genes in which variants are present in both children but there are no variants at all in one of the parents, as in this picture?

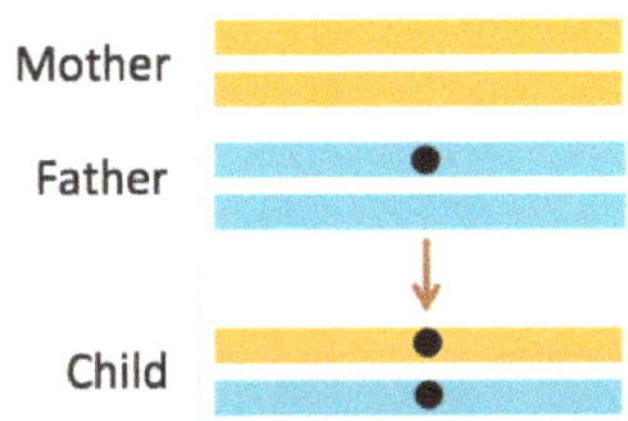

A scenario we might have overlooked—one parent appears not to have the variant at all.

On a hunch, we instruct our computer to wade through our millions of variants and find any genes we might have missed earlier—genes with variants in both genome versions in the children, but in *one or none* of the genome versions, in each parent. We wait, hoping for something new to change the face of this investigation. The computer produces its new list, which is much the same as the previous one. There is one notable addition, though—right at the top is a new candidate gene: *MMP21*.

Unfortunately, even less is known about *MMP21* than about *BCL9L*. There are no reports of any variants causing disease in humans. There is little or no knowledge on what the gene does. Except for one remarkable fact: hot off the press, Cecilia Lo's list, generated from her group's marathon effort of inducing variants in mice, has *MMP21* in it!

MMP21: No Variant in the Mother?

In both *Raj* and *Taj*, a single character T in the *MMP21* gene is deleted. As extra characters and deleted characters always do unless they come in multiples of three, this deletion causes a frameshift in the recipe. Character triplets in the recipe change dramatically as a result, just as they did in *Tara*'s case of heart failure in the third story.

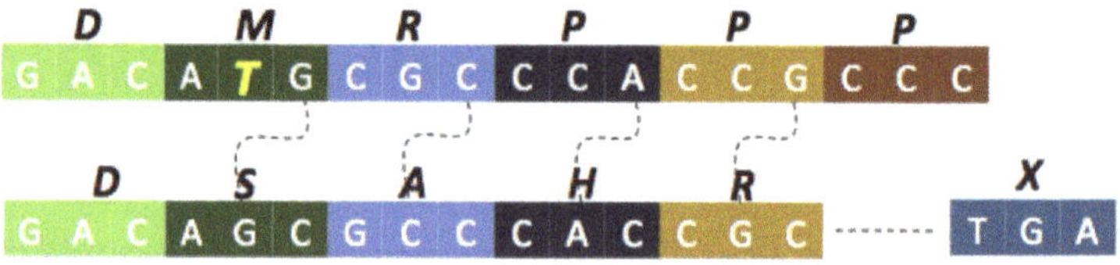

The reference sequence on top, and Raj's genome at the bottom, with the missing T highlighted in the reference. The recipe changes completely as illustrated by the amino acids in bold italics.

As happens often with frameshifts, the *Stop* triplet comes calling soon enough, thereby halting the gene recipe midway in its tracks, specifically at amino acid 175 instead of the usual 569. In both *Raj* and *Taj*, both versions of *MMP21* are so truncated. Their father has this recipe truncation in only one gene version. Strangely, their mother has no truncation at all; both her gene versions appear to be perfectly fine.

This is truly strange. If the mother doesn't have this variant at all, then how did both children acquire this variant in both their gene versions? It is very unlikely that both children acquired the variant *de novo*; that would be too much of a coincidence. What could be a possible explanation?

When no explanation suggests itself, Holmes's claim provides a prescription, the right one in this case: *either we are wrong about the mother, or we are wrong about the two children.* How did we conclude that the children had the *MMP21* variant in both gene versions, while the mother had it in neither version? We carefully re-examine any assumptions we made in this process, starting from how genome sequencing works.

Remember, when *Raj*'s genome was sequenced, we obtained little pieces of his genome called reads, each only about 100 characters long. We then put these pieces back together, much like we would assemble a jigsaw puzzle, using the reference sequence as a guide. *Raj*'s genome is not identical to this reference sequence but is very similar. So, a read from *Raj*'s genome must be present somewhere in the reference sequence as well, either in identical or in very slightly altered form. Accordingly, we hunted for each read in the reference sequence, allowing for a few character modifications, additions and drops. We then placed each read at its respective matching position, and obtained the following picture.

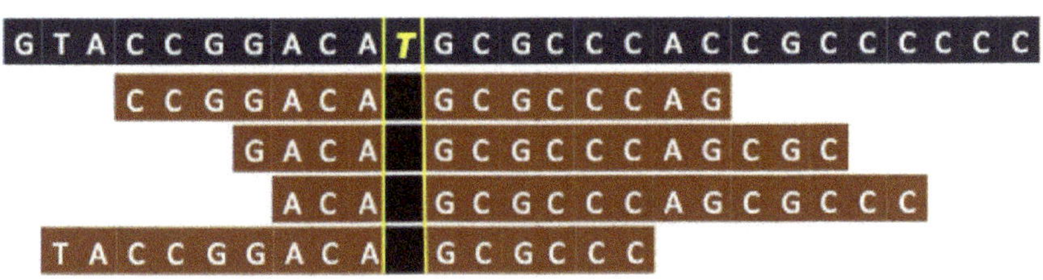

Reads in Raj and Taj with the alterations in black, juxtaposed against the reference sequence, which appears on top.

We noticed from this picture that *every* read in the children was missing a T. That missing T is represented by a dark blank, introduced so the remaining characters in these reads are in register with the reference. Accordingly, we concluded that the children have this variant in both gene versions. After all, had even one of the gene versions retained this T, we would have seen at least a few reads with the T. Instead, every single one of the reads we see here lacks the T.

Indeed, the corresponding picture in their father appears below and drives home the contrast.

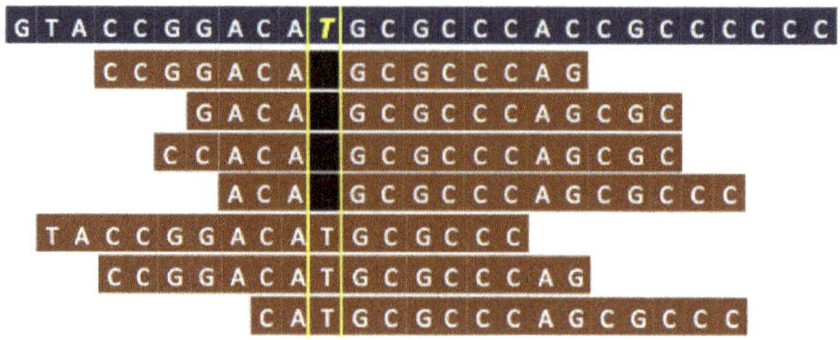

Reads in the father.

Here, roughly half the reads are missing the T. Since one expects roughly half the reads to come from each gene version, we concluded that the father is missing the T in only one gene version. And the mother? Her picture is shown below.

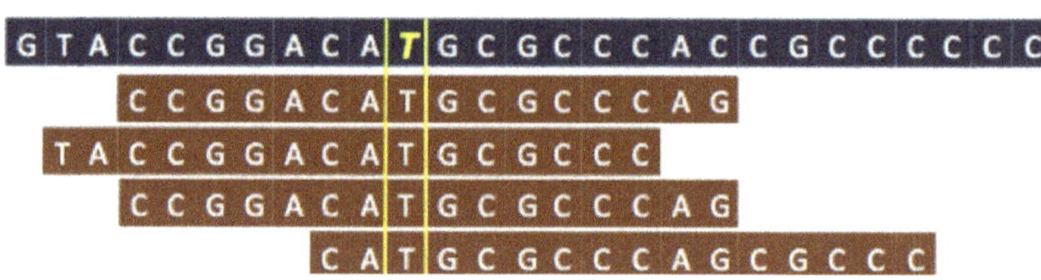

Reads in the mother.

All these reads have the T loud and clear. Accordingly, we concluded that the mother doesn't have this variant at all. Our conclusion appears solid. Then what are we missing?

A sharp eye will notice an interesting quirk in the three pictures above. *The number of reads in each of Raj, Taj, and their mother is fewer than what it is in their father.* Does this yield a clue? Why should it be so? Possibly, this is a sheer chance event that has no bearing on our quest.

Or, possibly... *Raj*, *Taj* and their mother are missing one version of *MMP21* altogether. Well, if not a complete version, then at least a large enough chunk around the T in question, as in this picture.

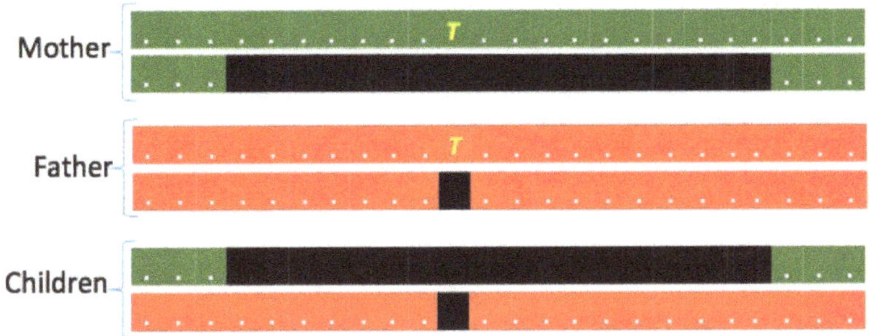

The two versions of MMP21 in each individual, with missing characters and hypothetical missing chunks shown in dark.

In this hypothetical scenario, both parents have only one fully intact gene version. It is the other gene version that holds the key to our puzzle. The father lacks a T in his other version. Both *Raj* and *Taj* have inherited this gene version from their father. The mother, on the other hand, lacks a much larger chunk in her other version. Unfortunately, both *Raj* and *Taj* have inherited this gene version from their mother.

So be it, but why is this a plausible scenario at all?

A little thought provides the answer: the large missing chunk contributes no reads of interest at all. *All* reads in the children will therefore arise from the gene version they inherited from their father, with just the T missing. *All* of these reads will lack this T. That tricks us into believing that the T is missing from both gene versions, when it is truly missing only from one. We might have been taken in by this illusion all along.

Since only one gene version contributes all the reads in this hypothetical scenario, there will be fewer reads in the children and the mother than in the father. That would explain it. Of course, provided we can prove that there is indeed a missing chunk. Thus begins our hunt for this chunk.

The Hunt for the Missing Chunk

Imagine the mother's genome with a large chunk missing from one version of the *MMP21* gene. Do her reads carry any signs of this missing chunk? What should we look for in these reads? The following picture helps set the stage.

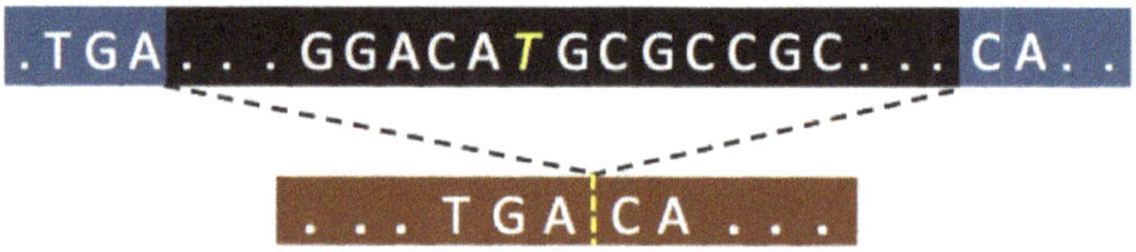

The reference sequence on top with the chunk missing from the mother highlighted in dark. A read from the mother's genome at the bottom jumps right across the chunk.

This picture shows the chunk missing from the mother's genome along with a read obtained from her genome. Remember how this read is obtained? Many instances of her genome are taken and chopped up into random fragments. Then the character sequence in each fragment is elucidated to obtain the corresponding read. What impact does the missing chunk have on these reads?

Well, some of these reads come from fragments that straddle the missing chunk. The read shown above is one such. The missing chunk is of course absent from this read. So, this read comprises characters to the left of the chunk followed by characters to its right. The boundary between these two sections, though marked in the picture, is purely imaginary; the read does not come with this boundary marked. That, in fact, is the problem we face when we hunt for this read in the reference sequence.

Remember, during this hunt, we do allow for slight alterations to be made to the read. A few character modifications, additions, or drops in the read qualify as slight alteration. For instance, in an earlier picture, we added a dark blank to reads from the father's genome to account for the missing T. However, our hypothesis here is that a much larger chunk is missing from the mother's reads. So, a few alterations will just not do; we need to insert a large number of dark blanks in these reads to stand any chance of obtaining a match in the reference sequence. Since we did not allow ourselves this luxury, our hunt for such reads would have failed to find a match in the reference sequence.

Of the billion or so reads we generated from the mother, it is typical for few tens of millions to not find any match in the reference sequence, for various reasons. These reads are parked aside and ignored thereafter. Hidden among these reads might just be the handful we need—the handful that straddle the large missing chunk, thus providing us evidence of its existence. How can we find these few needles in such a massive haystack?

To this end, we must take each read from this haystack and then launch a hunt for this read in the reference genome. This time, slight alteration is not sufficient; we have to allow for many more alterations. We don't even know how many, for we have no idea how long our hypothesized missing chunk is. And we have no idea where these alterations must be made within each read—toward the beginning, or the middle, or the end? How then do we hunt for such reads in the reference?

Read Hunting with Few Alterations

We begin by reviewing how we hunt for a read in the reference sequence allowing only a few alterations, with the hope that doing so will yield an opening toward the search for the missing chunk in *Raj*'s and *Taj*'s mother. These so-called *computer algorithms* can seem quite abstract, so feel free to quickly skim through what follows, or even skip it altogether in favor of the next section.

Consider one read, about 100 characters long, while noting that we have a billion such reads to handle. So, we better be quick with this read. However, this is easier said than done because the reference sequence is huge—three billion characters long. It takes way too long to check every one of these three billion positions explicitly and verify whether the read matches there with a few alterations or not. Fortunately, though we won't dwell on this, there are clever ways to quickly narrow these three billion positions down to a handful.[67,68] Each of these positions can then be examined to verify if the read indeed matches there with a few alterations. How this verification is done is best illustrated with an example.

The picture below shows a read GTCCGT and one position in the reference sequence starting with the characters GACAC. The read does match here with three alterations, as shown.

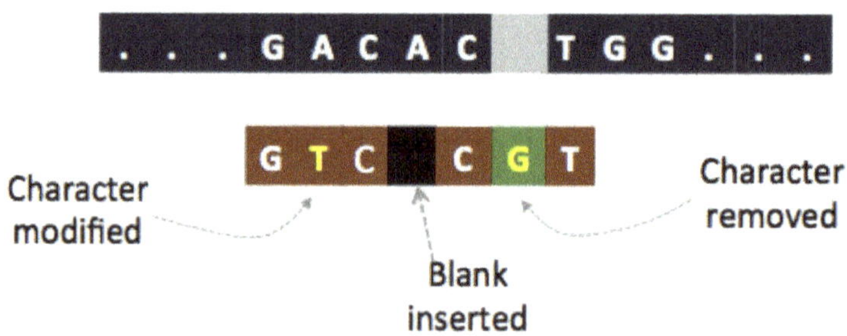

One combination of three alterations to the read GTCCGT that makes it match the reference sequence at the position starting with GACAC.

Arriving upon this combination of three alterations in not easy though. One may have to try many combinations to find one that works. Millions of combinations actually, if you consider even up to four alterations in a read of length 100. That is clearly too many to wade through before we know if four alterations will do or not. Is there a faster way?

A picture, they say, is worth a thousand words. In that spirit, the picture below concisely captures every one of the myriad ways to make alterations.

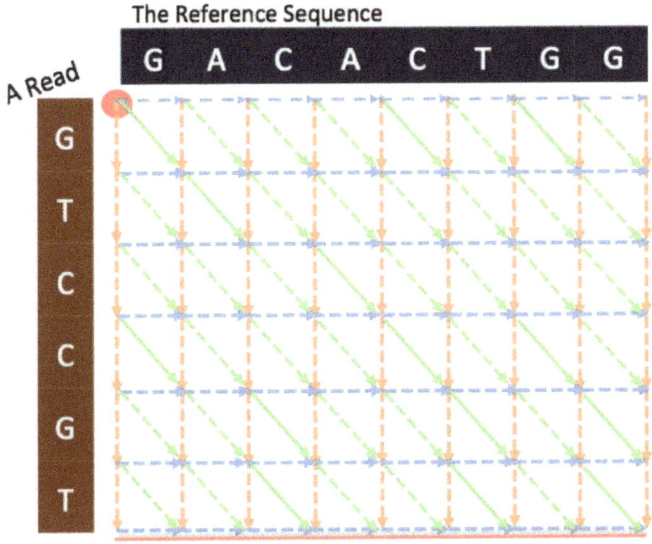

All possible alterations in one picture.

In all likelihood, you are stumped at this point. That this picture captures all possible ways to make alterations to the read is far from obvious. The key lies in the following fact: every distinct combination of alterations in the read corresponds to a path in this grid from the large dot at the top-left to any grid-point in the bottom row.

To let this sink in, try a few ways to walk through the grid from the large dot down to the bottom row. At each step, you have three choices: you can go down, go right, or go down and right diagonally. When you go down, imagine you are removing a character from the read. When you go right, imagine you are adding a blank to the read. When you go diagonally, imagine you are switching one character for another. Of course, there are a myriad ways to do this. Here is a specific example for the combination of three alterations depicted earlier.

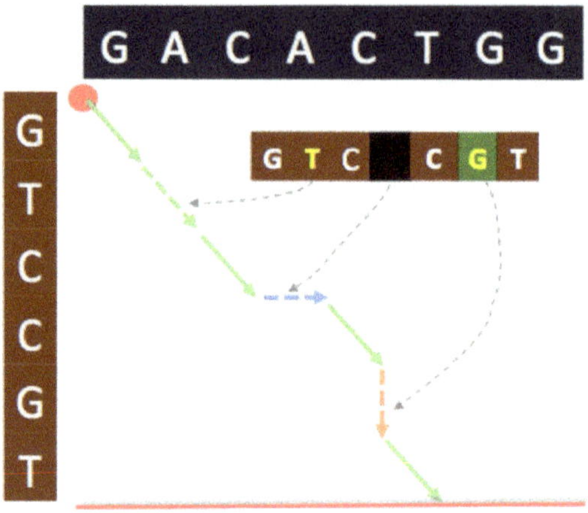

A path corresponding to the combination of three alterations for this read.

So, our task of identifying the right combination of alterations boils down to identifying the right path through this grid. If you haven't noticed yet, some of the lines in this grid are solid and some are dotted. It turns out that the number of dotted lines in a path indicates the number of alterations. So, our task boils down to finding the path with the *fewest* dotted lines.

Of course, since we limit ourselves to a few alterations, say four, our paths are allowed at most four dotted lines. This means our paths can take horizontal or vertical lines at most four times, for these lines are *always* dotted. Therefore, the path we seek must hug the diagonal for the most part, staying close to it within a narrow band. Since this band is narrow, one isn't left with that many options. So, the procedure to find a path with the fewest dotted lines through this band becomes super quick.

Of course, this method of hunting for a read in the reference sequence is super quick only if you allow very few alterations. Our hypothesized missing chunk is expected to be much longer, thus requiring many more alterations. Our band becomes much wider then, and the procedure becomes impossibly slow. To top it all, we don't even know how long the chunk is. What do we do then?

Read Hunting with Many Alterations

To identify this missing chunk from the mother's genome, we have to find a faster way to process the tens of millions of her reads which haven't yet found their matches in the reference sequence. Hopefully, a few of these reads straddle the missing chunk—they start to its left and then seamlessly continue to its right, blissfully unaware that several characters lie in between in the reference sequence. And this suggests an obvious line of attack.

What if we break a read into two parts; a prefix and a suffix? Each part by itself will hopefully find its match in the reference sequence with just a few alterations. The prefix will find its match immediately to the left of the chunk, and the suffix will find its match immediately to its right. Of course, the missing chunk might be long, so the prefix and the suffix might find their respective matches in the reference sequence far away from one another. Nevertheless, once we find these sandwiching matches, the missing chunk will become quite apparent—it is the chunk that lies in between these matches in the reference sequence.

Accordingly, we just split each read into two parts and hunt for each part separately, allowing slight alterations in each case, as we did earlier. And we repeat this for each of the tens of millions of reads in our haystack. As simple as that. Unfortunately, there is a small catch still.

There are 99 different ways to split a read comprising 100 characters into two parts: we could split after the first character, or after the second, and so on. How do we find the right split?

One can almost imagine Dr. House from the American medical drama, *House*, barking at us: *try ALL of these.* Of course, that is too laborious, and unnecessary, for there is a quicker way. We attempt to find the best path through the grid we discussed earlier, allowing just a few dotted lines, and knowing fully well that our effort is unlikely to yield such a path if this read straddles the long missing chunk. Nevertheless, we try to go as far down the grid as possible, using just a few dotted lines, before we are forced to employ many more dotted lines to plough our way further. The prefix-suffix boundary must lie somewhere here, and we split the read accordingly.

So the tens of millions of reads in our haystack are set rolling on this algorithm—entrusted to a computer, of course. Once it finishes, which it does quite quickly, we scan the reference sequence for positions where prefixes and suffixes of several reads stack up. And we look to see if these two stacks sandwich any part of the *MMP21* gene. And what do we find?

The Missing Chunk, Finally

Staring at us finally is the answer to our riddle of *Raj*'s and *Taj*'s peculiar architectural plan—sharp stacks of read prefixes and read suffixes that sandwich the mother's first three *MMP21* exons, signaling the missing chunk as in the picture below.

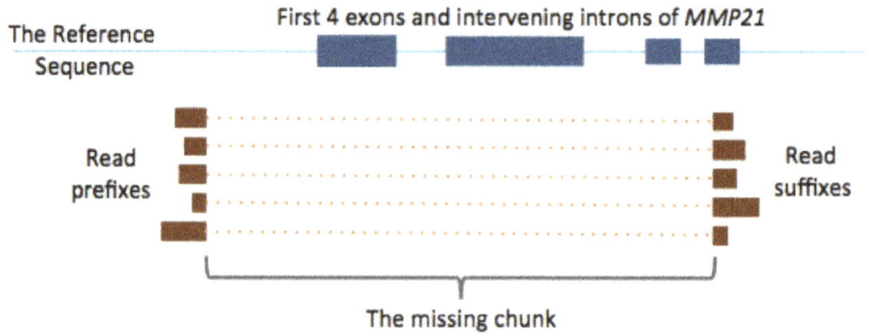

Prefixes and suffixes from many reads stacked up, sandwiching the missing chunk.

The missing chunk, obvious from this picture, starts about 2000 characters before the start of the gene recipe. This is no-man's land—there are no genes or exons here. It was fortunate we sequenced the whole genome; had we sequenced only the exons, there is a chance we might have missed it altogether. The missing chunk then continues through the first three exons, finally terminating in the middle of the fourth exon.

The chunk comprises 5925 characters; all these characters are missing from one gene version in the mother, rendering it dysfunctional. Nevertheless, the mother's anatomy remains normal because it is rescued by the other unaffected gene version.

We repeat the above procedure on *Raj* and *Taj* to verify that both inherited the abnormal gene version from their mother. As expected, stacks of read prefixes and suffixes sandwiching the missing chunk appear yet again. Both children have indeed inherited the missing chunk from their mother, making that version of their *MMP21* gene dysfunctional. They have also inherited the missing T from their father, making the other version dysfunctional as well. Taken together, *Raj* and *Taj* have no functioning version of the *MMP21* recipe at all. And that is the likely cause of the odd placement of their organs.

Wrapping Up

Raj's and *Taj*'s rather unusual case has led to the identification of a new gene *MMP21* for heterotaxy—a phenomenon where some but not all organs appear out of place, accompanied often by heart malformation. Problematic variants in *MMP21* have never been noticed hitherto in any human patient and little is known about the gene's function. Yet, we have strong evidence that *MMP21* is indeed the cause of *Raj*'s and *Taj*'s condition.

It is thanks to the marathon efforts of Cecilia Lo and her group that closure in this case became possible. Their screening of thousands of mice to identify one with an *MMP21* variant was key to our efforts. The initial *MMP21* variant they found was a missense variant: a change from *Tryptophan* to *Leucine* at amino acid 177. Subsequently, experiments in mice and in zebrafish have both confirmed that complete loss of the *MMP21* gene, as is effectively the case in *Raj* and *Taj*, leads to heterotaxy and heart malformation.[69]

However, we don't yet know how variants in *MMP21* perturb the usual process of breaking left–right symmetry, resulting in this odd placement of organs. Experiments suggest that recipe execution from *MMP21* is apparent only in a 10.5-day-old mouse embryo.[70] Since moving cilia break left–right symmetry between days 7.5 and 8.5, *MMP21* very likely plays its role in the subsequent cascade of events and not in the formation or functioning of the cilia. Only further experiments will uncover its exact role.

Raj and *Taj* have both needed heart surgery for survival. Their parents now know the exact reason why two of their children were born with unusual anatomical features, while they themselves, and their other children, have the standard anatomical plan. This information could be invaluable should they choose to have another child, as we saw in the story of *Rom* and *Som* earlier.

It did take a while to get to the bottom of this case. The answer lay buried all along under a gargantuan haystack of data. Fishing it out required a most unlikely combination: very careful sleuthing, some clever computer algorithms, and a marathon effort on screening mice.

Story 6

The Blood Can't Carry

Now we begin our story of *Sita*, a two-year-old girl who appeared unusually pale and malnourished. She was often short of breath, and seemed far less energetic than others her age. She also fell sick frequently. Her liver and spleen were enlarged, and her red blood cells appeared pale and oddly shaped under a microscope, as in this picture below. Her doctor had no trouble diagnosing her condition.

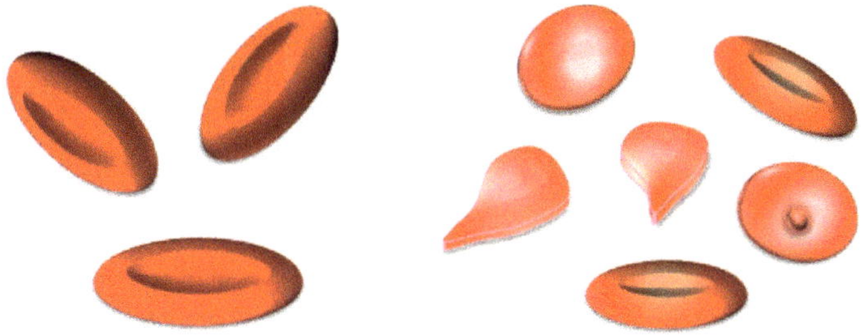

Normal red blood cells on the left, and smaller, paler and sometimes oddly-shaped red blood cells on the right. Uncharacteristically, some even have a nucleus.

The redness of red blood cells comes from a very familiar molecule called *hemoglobin*, which is instrumental in carrying oxygen from the lungs to the rest of the body. The paleness of these cells in *Sita* indicated that she had very low levels of functional hemoglobin.

Consequently, her blood could not carry and deliver oxygen in sufficient amounts, a condition called *Beta-Thalassemia Major* (or just thalassemia, for short). Serious medical intervention then became a necessity.

The immediate fix for thalassemia is a blood transfusion. Accordingly, blood from a willing donor was transfused into *Sita*. This provided her with a supply of red blood cells carrying good hemoglobin, thus allaying her problem. The problem did not go away though. Red blood cells live for just 120 days on average, after which they are destroyed. *Sita*'s body had to generate new red blood cells to compensate for these dying cells. But her body could only generate red blood cells with defective hemoglobin. So, she needed not one, but many blood transfusions, once every month or so. Indeed *Sita*, now three and a half, had already undergone several such transfusions.

What quirk in *Sita*'s genome had rendered her hemoglobin incapable of carrying its usual load of oxygen? How did she survive for several months inside her mother's womb, and then several months after birth, with this handicap? Would she be able to manage lifelong with transfusions? Was there a cure on the horizon?

The Oxygen Carrier

We begin our search for answers by understanding hemoglobin better. What strikes us immediately is its rather intricate design. Like a sponge that can soak up and let loose large amounts of water, hemoglobin is designed to absorb and then deliver large amounts of oxygen rapidly.

The recipe for the creation of hemoglobin is written in the genome, not in a single gene, but across two different genes—*HBA* and *HBB*. Two protein molecules created using the *HBA* recipe and two created using the *HBB* recipe come together to create a single entity. The four parts of this entity are called *globins*. This is not hemoglobin yet, mind you; that needs one more step, for globins, by themselves, are incapable of carrying oxygen.

These globins need an additional attachment called *heme* to make hemoglobin. There are four such attachments, one wedged deep inside each of the four globins. Unlike the globins, the recipe for heme is not directly coded in any gene. Rather, the body makes heme from raw materials which we obtain from our environment. Together, four globins and the four hemes make a single molecule of hemoglobin, shown below.

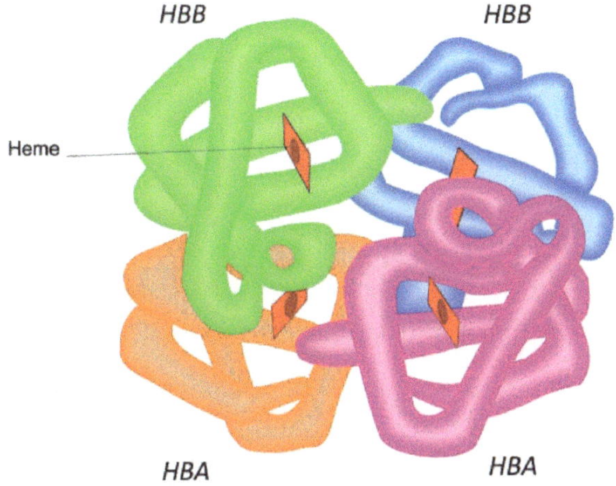

Hemoglobin, with its four globins and four hemes.

Fascinatingly, each red blood cell in our blood contains millions of hemoglobin molecules. And each of us has, possibly, trillions of red blood cells. That is a lot of hemoglobin!

At the heart of each heme is the scene of action—an *iron* atom. This is where oxygen attaches itself. In the lungs, where oxygen is abundant, much is soaked up by this iron, giving oxygenated blood its brilliant red color. Red blood cells carrying this oxygenated hemoglobin then travel through the bloodstream to other parts of the body. The small amount of oxygen dissolved in blood gets quickly exhausted here. The resulting drop in oxygen concentration serves to squeeze the sponge, forcing oxygen out from heme and making it available to the various tissues.

Coming back now to rapid soak-up and delivery of oxygen—this is enabled by an intricate structural quirk in hemoglobin. To appreciate this quirk, imagine hemoglobin with no oxygen attached to any of its four heme attachments. In this form, each heme molecule has a pyramid-like shape, as shown below, and shows relatively little interest in oxygen.

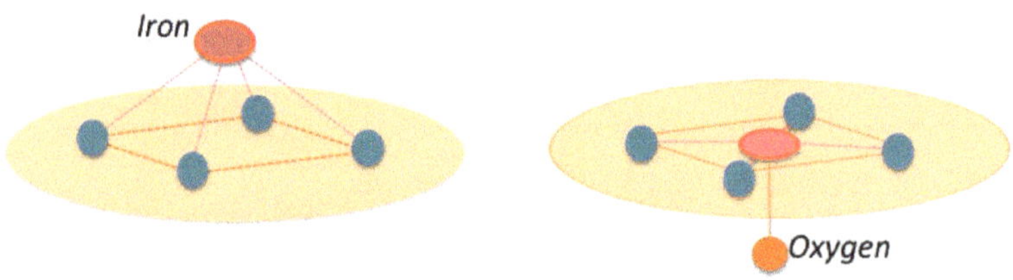

Heme—without, and with, attached oxygen.

When oxygen does attach to the iron in any of the four hemes, the structure of that heme flattens out, as shown above. This has a cascading effect, forcing structural adjustments in the three other globins making their hemes much more interested in oxygen. As a consequence, oxygen attaches to the other three hemes quickly, thus allowing rapid soak-up. Likewise, detachment of oxygen from one heme expedites detachment from the other three hemes as well, thus allowing rapid delivery.

Such intricate structural orchestration is not without its accidental quirks. Carbon monoxide attaches to the iron in heme even better than oxygen. In fact, 250 times better! Thankfully, the amount of oxygen in our atmosphere is a million times that of carbon monoxide. If, for some reason, the amount of oxygen were to reduce to only 250 times that of carbon monoxide, 50% of our hemoglobin would be busy ferrying carbon monoxide instead of oxygen, fatally starving our cells of necessary oxygen.

Hopefully, there is no immediate threat of carbon monoxide levels rising to anywhere near such dangerous levels, except perhaps if one is caught in a fire. The greater threat, at least for the moment, lies in various altered gene recipes that prevent our oxygen sponge from functioning as intended, as seemed to be the case with *Sita*.

Sticky Hemoglobin

This leads to the next question—how do genomic variants alter hemoglobin? The expectation of course is that they reduce the snugness with which heme fits into globin, thus crippling the intricate orchestration needed for soaking-up and dispensing oxygen rapidly. Surprisingly, their impact can go well beyond.

An example is a flip from character A to character T in the *HBB* gene. This variant results in amino acid *Valine* at position 6 instead of the usual *Glutamic Acid*. The snugness of heme inside globin remains unaffected by this change. Unfortunately, something far more unexpected happens: hemoglobin molecules, which are tightly packed in red blood cells, start sticking to each other at amino acid 6 and form chains. These chains are rigid, so they contort red blood cells from their usual round shape into a sickle-shape, as in this picture.

Sickling of red blood cells caused by hemoglobin molecules sticking together.

Well, so what? The diameter of a red blood cell is about 7

microns (a micron is a millionth of a meter). Many arteries and veins are wide enough to allow easy passage for these cells. Not all though. Leading off these larger blood vessels is an elaborate network of narrower blood vessels called *capillaries*, which take blood to every nook and corner of the body. Many of these capillaries have diameters even smaller than that of a red blood cell. As a result, red blood cells must literally squeeze their way through these capillaries. This needs a lot of flexibility, which is lost with the rigid sickle-shape.

So, sickle-shaped red blood cells cannot easily squeeze through these capillaries. This blocks capillaries on the one hand, leading to pain and organ damage. On the other hand, red blood cells experience increased wear and tear, reducing their lifespans markedly—from 120 days to about 20 days. The result is a shortage of red blood cells, and therefore a shortage of hemoglobin that can deliver oxygen.

For obvious reasons, this disease is called *sickle cell anemia*. In certain tropical regions of Africa and India, more than 10% of us carry the aberrant *Valine* in one of our two *HBB* gene versions. The modest sluggishness that it introduces in our red blood cells does not cause serious problems; on the contrary, it appears to have a most unexpected protective effect—protection from malaria. However, those who carry this variant in both gene versions are affected seriously. Several hundreds of thousands of such new patients are reported every year.

Sita was not one of these patients though—her red blood cells were not sickle-shaped. Chances are she had a different type of recipe alteration in her genome. It is far from obvious what that alteration might be. And so begins our hunt for this alteration.

Deficient Hemoglobin

Starting off, our hunt needs some more familiarity with the *HBA* and *HBB* genes. Hemoglobin is jointly created by both genes, so recipe alterations in either could be problematic. Fortunately, nature has provided us with an extra level of protection—each of us has not just two, but four versions of *HBA*, and two versions of *HBB*.

Four versions of HBA, and two of HBB.

Remember, both *HBA* and *HBB* contribute equally to hemoglobin. Yet, we have more versions of *HBA* than *HBB* in our genome. Our cells do a counter-balancing act by executing the *HBA* recipe at roughly half the rate of *HBB*, so equal amounts are generated from both genes. However, altered recipes sometimes upset this fine balancing act.

For instance, recipe alterations could render some of the *HBA* gene versions dysfunctional. Fortunately, this has little impact if only one or two gene versions are so affected. Luck runs out when three versions are affected. The body can generate only small amounts of regular hemoglobin from the sole remaining version, compromising oxygen delivery seriously. The consequences can range from mild anemia to requiring occasional blood transfusions. Of course, when all four versions of *HBA* are lost, the impact is drastic—death occurs, before or shortly after birth.

Similarly, altered recipes in *HBB* that lead to loss of one version cause very mild impact. As you would suspect, things get more serious when both versions are lost. There is no *HBB* protein left to create hemoglobin with. Some oxygen is still delivered, either dissolved in the blood, or carried by hemoglobin-like combinations of *HBA* and genes other than *HBB*, but not nearly enough. What follows is an even more serious condition than caused by the loss of three versions of *HBA*. Periodic blood transfusions become absolutely essential.

You might wonder why the loss of both versions of *HBB* is not as immediately lethal as the loss of all four versions of *HBA*. It turns out that without *HBB*, a small amount of hemoglobin can still be created by combining *HBA* with other genes similar to *HBA* and

HBB. However, without *HBA*, no hemoglobin can be created at all.

Of course, loss of all four versions of *HBA* is a very unlikely event. Even the loss of three of four versions has very low odds. In comparison, the loss of one or even two versions of *HBB* is far more common. Studies on children in Mumbai and New Delhi indicate that one in 25 is a carrier of a problematic variant in one version of *HBB*[71]—roughly one in each classroom!

Of course, most of these children are very mildly affected. However, at even one child per classroom carrying a problematic variant, it is no surprise that several children are born with problematic variants in both gene versions, leading to beta-thalassemia. *Sita* was one such child. What alteration in her *HBB* recipe compromised her hemoglobin so severely?

The Search for the Genomic Variant

To answer this question, we need to sequence *Sita*'s genome. In the previous stories, we cast our net wide and sequenced all exons from all 20,000 or so genes in our subjects. In *Sita*'s case, that might be an overkill, for her problems are very likely to be specific to the *HBB* gene. Can we sequence this gene alone and avoid the rather elaborate process of casting a wider net?

The *HBB* gene makes things particularly convenient in this regard. It is all of 1605 characters long, with just three exons and two introns. This is very small as genes go, the average gene being about 10 times longer.

The HBB gene: three exons and two intervening introns.

It is indeed easier to sequence such a short gene directly rather than cast a wider net. So, we sequence this gene and identify all variants—characters where *Sita* differs from a supposedly healthy reference sequence. Of course, our focus is on the three exons of *HBB*, for gene recipes are encoded in these exons; the intervening introns are mere breaks in the recipe. There must be a problematic variant in these exons, given *Sita*'s symptoms are a perfect match for beta-thalassemia. However, at first sight, there isn't one. We check and recheck. But no variant here.

In the Introns?

Where was the elusive variant responsible for *Sita*'s condition hiding? It wasn't in the three exons of *HBB*. Where else could it be? Could it be in the introns?

Introns, at least at first sight, are mere breaks in the recipe carried by the exons. Surely, variants in these introns cannot affect the recipe in any way. A wave of skepticism passes through our mind. Then, a moment of thought raises a fundamental question.

True, introns do not carry the gene recipe. So, when the recipe is executed and protein molecules created from it, the introns must be skipped over—or, to use the right terminology, *spliced out*. But introns can be very long, a few thousand characters on average, even going up to tens of thousands of characters in some cases,[72] which is much longer than most exons. How do our cells know to stop precisely at the end of an exon, and then jump over a huge intron, all the way to the start of the next exon?

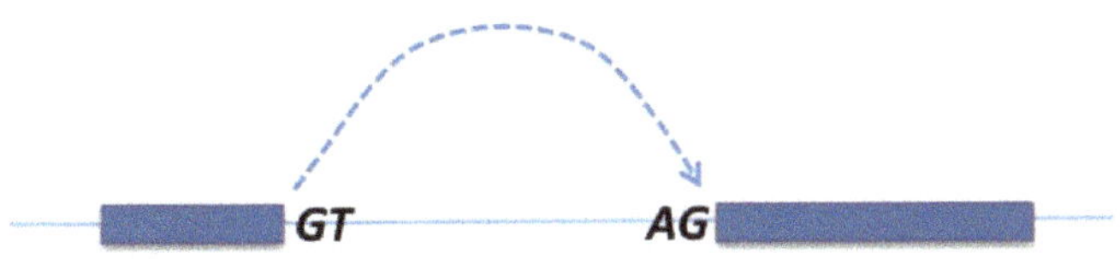

Recipe execution jumps from GT to AG, thus splicing out the intron.

The answer lies in special character sequences that serve as boundary markers between the exons and introns. Invariably, the beginning of an intron is signaled by the pair of characters, GT, and the end is signaled by the pair of characters, AG, as in the picture above.

When the *HBB* recipe is executed, these markers are used to delineate and remove the introns. This brings the three exons together. The recipe in these exons is then partitioned into triplets and each triplet is converted into an amino acid, thus yielding the *HBB* protein.

And that sets us thinking—what happens when a genomic variant changes one of these boundary markers? Wouldn't the splicing jumps go haywire? For instance, if the boundary marker GT changes to, say, GG, then wouldn't the picture look like the one below?

Recipe execution continues into the intron when GT becomes GG.

This picture shows the recipe simply continuing into the intron rather than pausing and jumping over—willy-nilly, the intron seems to have now become part of the recipe. Of course, this accidental intronic recipe is quite different from the recipe in the exons. And, very likely, as in our previous stories, this intronic recipe encounters a premature *Stop* instruction and is severely truncated—a rather drastic outcome of a simple boundary marker change from GT to GG.

A similar situation seems to arise if the boundary marker AG changes, say, to AT. Recipe execution launches its jump at the GT but may not land where it ought to; chances are that it lands somewhere else, maybe much farther away, splicing out a whole exon along with both introns in one shot, as in the picture below. Again, the recipe that results is very different from the original.

Recipe execution skips both introns, along with the exon in between, when AG becomes AT.

Sporadic literature on this topic yields a word of caution though: it is not always easy to predict how exactly a recipe changes when a boundary marker is modified. Indeed, we will see a rather unexpected example in a later story. However, there is little doubt that the change is dramatic, and, in all likelihood, cripples the recipe seriously.

Maybe this could be the source of *Sita*'s problem. With renewed hope, we look for variants in the exon boundary markers in *Sita*'s *HBB* gene. Unfortunately, we draw a blank again. All her GTs and AGs are fully intact.

Long Hops and Short Steps

Where else could *Sita*'s variant be hiding? With the hope that a more careful look might yield a clue, we stare at the sequence of characters in the second of the two *HBB* introns. Not *Sita*'s sequence of characters, that'll come later, but rather, just the sequence in the average person. As expected, it begins with a GT and ends with an AG, as in the picture below.

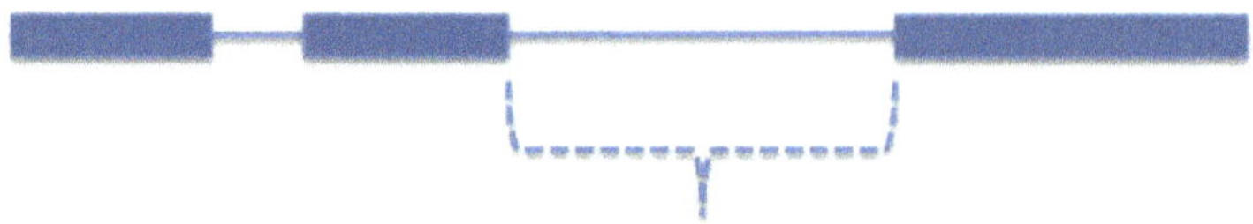

```
GTGAGTCTATGGGACGCTTGATGTTTTCTTTCCCCTTCTTTTCTATGGTT
AAGTTCATGTCATAGGAAGGGGATAAGTAACAGGGTACAGTTTAGAATGG
GAAACAGACGAATGATTGCATCAGTGTGGAAGTCTCAGGATCGttttagt
ttcttttatttgctgttcataacaattgttttcttttgtttaattcttgc
tttctttttttttcttctcCGCAATTTTTACTATTATACTTAATGCCTTA
ACATTGTGTATAACAAAAGGAAATATCTCTGAGATACATTAAGTAACTTA
AAAAAAAACTTTACACAGTCTGCCTAGTACATTACTATTTGGAATATATG
TGTGCTTATTTGCATATTCATAATCTCCCTACTTTATTTTCttttatttt
taattgatacataatcattatacatatttatgggttaaagtgtaatgttt
taatatgtgtacacatattgaccaaatcagggtaattTTGCATTTGTAAT
TTTAAAAAATGCTTTCTTCTTTTAATATACTTTTTTGTTTATCTTATTTC
TAATACTTTCCCTAATCTCTTTCTTTCAGGGCAATAATGATACAATGTAT
CATGCCTCTTTGCACCATTCTAAAGAATAACAGTGATAATTTCTGGGTTA
AGGCAATAGCAATATCTCTGCATATAAATATTTCTGCATATAAATTGTAA
CTGATGTAAGAGGTTTCATATTGCTAATAGCAGCTACAATCCAGCTACCA
TTCTGCTTTTATTTTATGGTTGGGATAAGGCTGGATTATTCTGAGTCCAA
GCTAGGCCCTTTTGCTAATCATGTTCATACCTCTTATCTTCCTCCCACAG
```

34 AGs in the second intron of HBB.

A question arises immediately. The AG at the end is not the only AG in this intron; there are as many as 33 other AGs. A splicing jump that takes off from the GT boundary marker at the very beginning must land specifically on the 34th AG for this intron to be spliced out correctly. What makes this AG special? And what guides our cells to this special AG, among several others before and after in the vast genomic ocean?

Investigations reveal that this honor belongs to an intermediate character called the *branch point.* This point breaks the splicing jump from GT to AG into two parts: a long hop, and a shorter step,[73] as in the picture below.

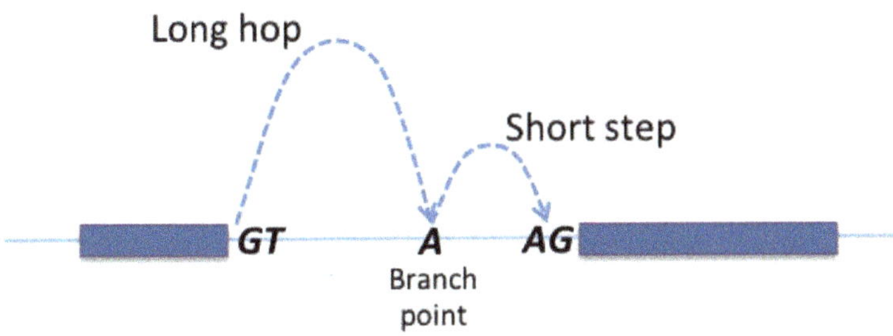

The long hop to the branch point, and then the short step.

That only begs the next question—where in the vast intronic wilderness does the branch point lie? How do our cells find this point?

A few facts about branch points catch our attention as we ponder over this question. First, the character at the branch point is most often an A.[74] That's a good start. However, not surprisingly, there are over 100 As to choose from in the second intron of *HBB*, which was depicted a little earlier. Which of these serves as the branch point? It turns out that only those that match the following pattern are candidates.

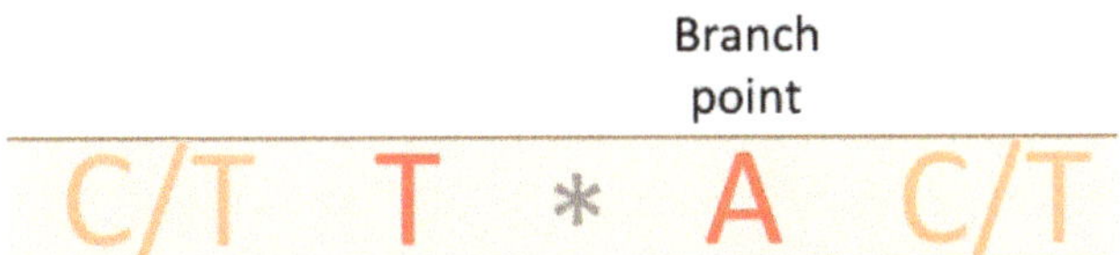

The pattern of characters around the branch point character A.

This pattern requires that the character two steps to the left of the A be a T. The character in between is allowed to be anything, hence the *. And the characters at either end must be C or T. This pattern is not an overwhelmingly strict requirement—some variations are allowed; in particular, occasionally As or Gs might appear at either end, hence the lighter hue for these characters.

Nevertheless, even treating this pattern as sacrosanct, there remain 27 candidate As in the second intron of *HBB*. Which of these is the true branch point?

We unravel the onion some more. A bit to the right of the branch point must lie a contiguous stretch of characters comprising predominantly Cs and Ts, as depicted below. The occasional A or G might appear here but Cs and Ts are predominant. This stretch must be about 20 characters long,[74] with the first few characters more likely to be Ts than Cs.

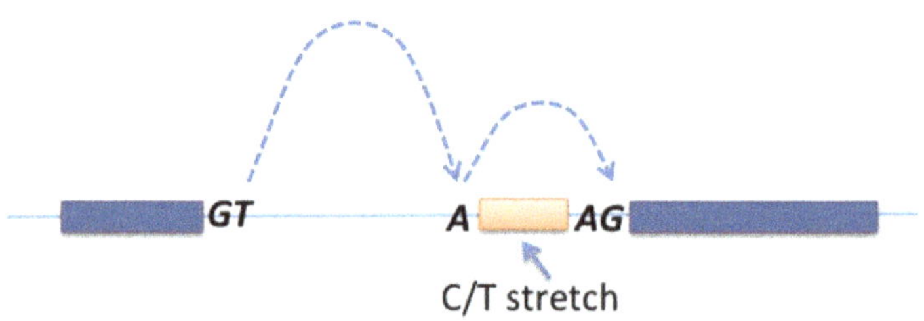

The C/T stretch between the branch point and the AG.

A scan of the characters in the second intron of *HBB* shows that only seven of the 27 branch point candidates have such an accompanying stretch. Our hunt for the branch point is down to these seven candidates now.

Then, it turns out that the branch point must typically be within 40 characters of the destination AG.[75] That's not surprising, for we did call it a short step. Only the very last of our seven candidates fits this bill this time. And that, finally, is our branch point, in all likelihood—the point that breaks the splicing jump into a long hop and a short step. Here is how the picture looks around this point.

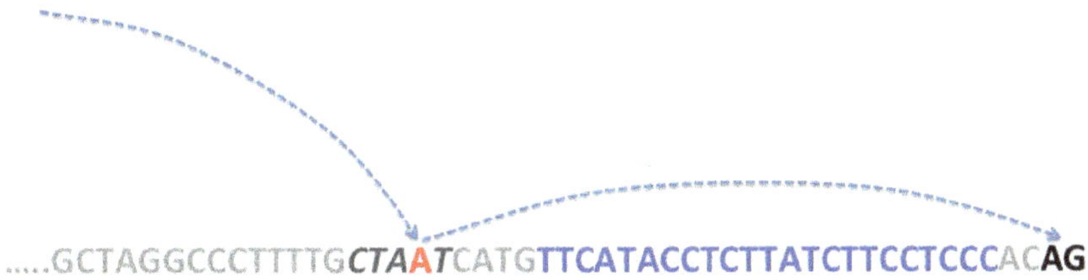

The likely branch point in intron 2 of HBB. The highlighted characters to the right of the branch point comprise the C/T stretch.

With this branch point at hand, and with some renewed hope, we surmise that a variant at this branch point might be the source of *Sita*'s problem. Such a variant might have confused the long hop, causing it to land at some other, unintended landing spot. The short step thereafter might have led to the wrong AG—not the desired AG at the end of the intron.

So, we look for variants in *Sita*'s genome at the branch points in both the introns of *HBB*. And we draw a blank yet again.

Where Hides the Variant?

Where else could the variant be hiding? It isn't in the exons. It isn't at the GT or the AG boundary markers. It isn't at the branch point. The sequence around this branch point is CTAAT, consistent with the branch point pattern depicted earlier; no variants here either. What remains?

There is an innocuous-looking variant buried within the C/T stretch in the second intron of *HBB*. What is usually a T seems to have become a G in both versions of the *HBB* gene in *Sita*, giving the C/T stretch an unwanted character. Only three of the 23 characters in this stretch are anything but C or T in the average person. But *Sita* has an extra G in this stretch, taking this number up to four. Here is how this extra G appears, right in the middle of the C/T stretch.

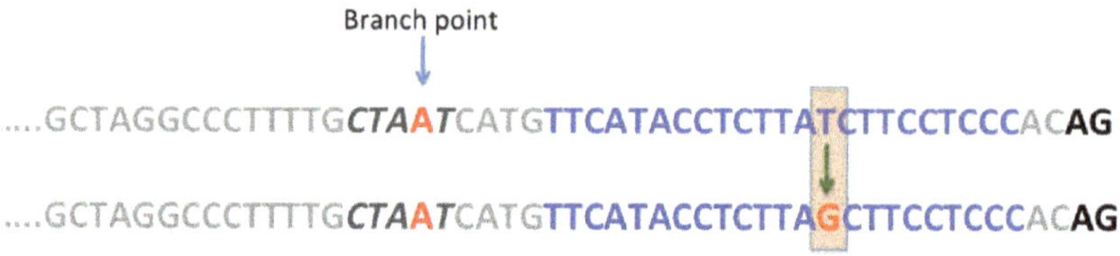

The reference sequence above, and Sita's sequence below, depicting the variant in the C/T stretch.

As shown in this picture, the average person has three A/Gs and 20 C/Ts in this stretch. *Sita*, in contrast, has four A/Gs and 19 C/Ts. Did this weaken her C/T stretch substantially? Some investigation reveals that is unlikely, thanks to the CTAAT pattern around the branch point; it matches the stipulated pattern perfectly, which compensates for a few violations in the C/T stretch.[76] On all counts, this variant looks truly innocuous.

We appear to be at a dead end. In desperation, we walk through the trail of events again. Introns in the *HBB* gene are spliced out during recipe execution. Splicing off an intron, say intron 2 of *HBB*, involves a long jump from the GT to the AG. This long jump occurs in two parts. A long hop first finds the branch point A, guided by the pattern of characters in its vicinity and by the C/T stretch. Then a small step takes the splicing process from the branch point to the next AG to its right. Presumably, the rightmost AG at the very end of intron 2.

Wait a minute! Which is the next AG to the right of the branch point? Usually, this is the AG at the very end of the intron. In *Sita*'s case, look carefully at the picture above. To her sheer misfortune, the T→G variant happens to be adjacent to an A. And this creates a brand new AG closer to the branch point! The long hop may well have found the branch point correctly, but only for it to be deluded at the very next step by this new AG. This AG entices the short step from the branch point toward itself and away from the intended AG, which lies 11 characters further to the right. The outcome: the splicing jump lands 11 characters too soon, as in this picture.

The splicing jump lands 11 characters too soon.

What happens next after this mis-landing? The recipe resumes from this landing point, forming character triplets, and converting each triplet to an amino acid. Since it resumes 11 characters too soon, it creates a few more amino acids in *Sita* than it would do in the average person. But that is not all.

Unfortunately, 11 is not a multiple of three, and therefore the triplets in the recipe change completely, as this picture shows, leading to a very different sequence of amino acids.

Who would have thought that a seemingly innocuous T→G variant buried inside an intron could so dramatically change the latter third of *Sita*'s *HBB* recipe.

No doubt, this is a rather serious change to the recipe. That too in both versions of the *HBB* gene. Is there even an outside chance that the hemoglobin obtained from this modified recipe functions as well as the original? If so, then *Sita*'s thalassemia would remain unexplained and our quest to find the underlying cause must continue. So, we investigate the effects of this modified recipe on the oxygen-carrying capability of hemoglobin.

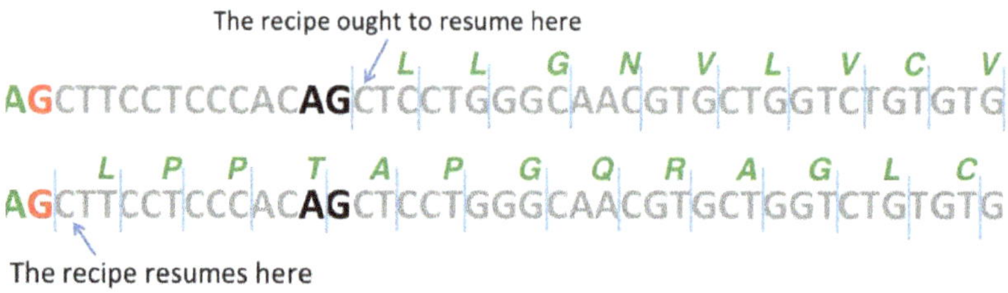

The normal recipe at the top, and Sita's dramatically different recipe at the bottom. Short forms of amino acids are shown above each triplet.

Does Heme Still Fit Snugly?

Remember the four globins in hemoglobin, and their heme attachments? Two of these globins come from the *HBB* recipe. *Sita*'s *HBB* recipe appears to be different from most others. Does heme still fit snugly into the two globins generated from this recipe?

The answer, of course, depends upon the shape of these globins. Here is how this shape is obtained. The execution of the *HBB* recipe creates a chain of amino acids. Interactions with each other and with surrounding water molecules now rotate bonds in this chain, causing it to fold. This folding continues until the amino acid atoms come to occupy certain equilibrium positions in space, with limited ability to move from these positions. This yields the characteristic globin shape.

Note this is not a random shape but a highly repeatable one—roughly the same atom positions result every time the *HBB* recipe is executed. It is as if the desired shape is written in the recipe itself. Of course, this shape has special capabilities: it fits heme snugly, and it adjusts itself when oxygen attaches to enable rapid absorption and delivery. We need to verify that the shape resulting from *Sita*'s modified recipe does not retain these special capabilities.

To this end, a picture of the shape might help. By shining X-rays on globin crystals and studying diffraction patterns, crystallographers have inferred precise positions of the thousands of atoms that comprise globin. However, depicting these positions individually is not easy, and the resulting clutter is too much for the eye. So, we imagine a ribbon as it winds its way through the various amino acid positions, as in the picture below. In particular, a ribbon made of metal that holds its shape, but can also be bent.

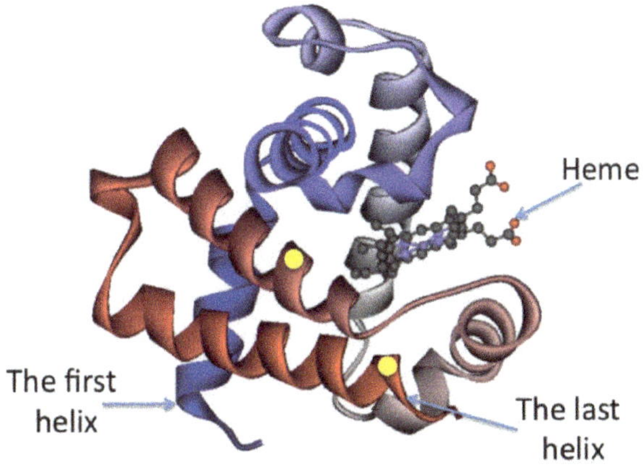

The shape obtained from a normal HBB recipe. Colors along the amino acid chain gradually change from one end to the other.

As you can see, some stretches of this ribbon have twisted themselves into spring-like helices. You should be able to count seven helices in all. These helices behave like rigid rods, which are so oriented that they wall off a central pocket from all sides. This pocket is tailor-made for the heme attachment to fit in.

You may think from this picture that heme is free to slide in and out of this pocket. That appearance is deceptive. Forces of attraction between the heme molecule and several amino acids in globin hold the heme in place and provide the required snugness.

Two of these, one in each of the last two helices, are amino acids 107 and 142, both marked with prominent dots in the picture above.

Now what happens with *Sita*'s globin, given its new unusual recipe? From amino acid 107 onwards, her recipe is dramatically different. Note the number 107—the two prominent dots in the picture above that provide anchor for heme have changed. Heme's snugness is possibly under threat now. How can we tell if the new set of amino acids in *Sita* continue to provide anchor to heme?

The only quick way is to use computer simulations to decipher the shape that results from this new recipe. These simulations attempt to mimic what actually happens in our cells as the *HBB* recipe is executed and the resulting amino acid chain folds into shape. Folding occurs because the various amino acids and other surrounding entities like water molecules exert electrochemical forces on each other. These forces cause atoms to change position. In their new positions, these atoms exert a new set of forces on each other. These new forces, in turn, cause further changes in position. And so on, repeatedly, until an equilibrium is reached. This whole cascade takes just microseconds inside a living cell.

We simulate exactly this process on a computer. However, simulating even a microsecond on a single computer can take weeks or even months. We use some tricks to expedite this computation, and relatively quickly get an approximate folded view of the last two helices, as in the next picture.

Contrast the tight, spring-like regularity of the helices in a normal person's globin to the rather floppy and disordered appearance of *Sita*'s helices in this picture. These rigid helices have started to come loose and the pocket they enclose is probably much bigger than it should be. As a result, heme is unlikely to fit snugly into this pocket anymore. At an extreme, this rather loosely folded structure might even be recognized by our cells as abnormal and destroyed.

If our computer simulations are correct, then the modified recipe in *Sita* provides every indication of compromising hemoglobin's ability to carry oxygen. Nevertheless, it may be wise to do an additional check and gather further corroborating evidence, just in case our simulations are not reflective of reality.

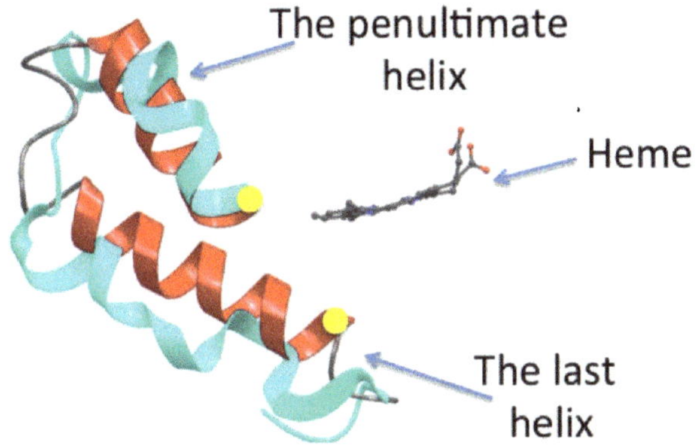

The last two helices in Sita juxtaposed with the normal picture.

Nailing the Variant

By way of an additional check, we scan scientific literature for some more evidence indicating the culpability of *Sita*'s T→G variant.

Studies on European and American patients seem to have no mention of this variant. Neither do studies on north Indian patients. This variant seems exceedingly rare in most parts of the world. So, we turn our eye to studies on patients from south India.

We find a study[77] that screened several thalassemia patients in the since-bifurcated South Indian state of Andhra Pradesh for genomic variants in the *HBB* gene. It identified six variants, each of which was present in multiple patients, and three additional variants, found in one person each. One of the latter three variants is exactly *Sita*'s T→G variant.

A more recent study,[78] this time in the neighboring south Indian state of Karnataka, also found this variant in two thalassemia patients. And yet another study,[79] an older one, found this variant in a patient of Indian origin. So, in addition to *Sita*, we know of four patients with this variant. On the other hand, none of the several tens of thousands of normal individuals we have scanned have this variant in either gene version. Surely, we have convincing proof of its culpability now.

Like most of these patients, *Sita* too was from south India. Maybe this variant had arisen somewhere in this region and had then passed down the generations. Still very rare, it usually appeared in only one version of the *HBB* gene in carriers of the disease. The unfortunate confluence of two defective versions, while very, very rare, did rear its head from time to time. *Sita*'s genome was one such confluence point. What could *Sita* do to keep the ill-effects of this variant at bay?

Hemoglobin from Donors

Sita's genome makes only defective hemoglobin, which cannot carry and deliver oxygen in sufficient amounts. Blood transfusions are the only immediate way to keep her healthy. Such transfusions are done quite routinely nowadays, and healthy and willing donors with matching blood groups are not hard to find. However, the challenge is that red blood cells have short life spans. *Sita*'s doctor might suggest removal of her spleen, an organ which accelerates the destruction of donor blood cells. Even then, *Sita* will need blood transfusions every month or so.

Such repeated transfusions will no doubt be cumbersome for *Sita*. They will also place her at an increased risk for blood-borne infections. Most importantly, though, they will expose her to a greater long-term risk.

Remember, each hemoglobin molecule has four heme attachments, and each heme carries an iron atom. Our body controls iron levels very tightly. Repeated infusions of iron-containing hemoglobin defy this control and excessive iron accumulates in the body as a result. High iron levels, in turn, lead to organ damage, particularly to the heart and the liver. *Sita* could face complications on this account. Treatment can mitigate some of these complications. Even then, *Sita* would be fortunate to live for a few decades.

Sita is still young and her family surely appreciates the reprieve provided by repeated blood transfusions. But the specter of complications and impending organ failure will no doubt weight heavily on their minds. Is there an alternative?

Replacing the Blood Genome

Here is an anecdote, which the internet attributes to Abirami Chidambaram of the Alaska State Scientific Crime Detection Laboratory.[80] It pertains to a case of sexual assault. Semen traces were collected from the scene of crime. The genome of the criminal was extracted from these traces. It was then matched against a genomic database of various criminals, constructed from their blood samples.

For purposes of this matching, it is too expensive, and a huge overkill, to compare the entire genome; it suffices to consider only a few carefully chosen genomic regions. These so called *repeat regions* carry sets of repeating genomic characters—for instance, ACGACTACGACTACGACTACGACT.., where the unit ACGACT repeats several times. These regions are chosen because of high variability from person to person—it is rare that two individuals have the same number of repeats. For instance, you might have five repeats of ACGACT, while I might have eight. By counting repeats at a few such places in the genome, an individual can be identified uniquely.

Coming back to the point, when the genome of the criminal in this case was matched against known criminals in the police database, there was a match. Mysteriously, a match with a person locked up in prison!

This was baffling. Did the prisoner quietly escape, commit the crime, and then return himself to jail? That was very unlikely. Further detective spadework yielded yet another match. What's more, the two matches were brothers!

Of course, brothers share more of their genomic sequences than two unrelated individuals do. Yet, even brothers have genomes that differ enough that the chances of counts matching at all repeat regions are very low. What could explain this mystery?

Another genomic test was performed to compare the genomes of the two brothers with the semen sample taken from the crime scene. This time, a cheek swab was taken instead of a blood sample. A cheek swab uses cotton to scrape off cheek cells from the inside of the cheek; the genome is then extracted from these cells. This time around, only the genome of the prisoner's brother matched the semen sample, and the genome of the prisoner did not. This nailed the prisoner's brother conclusively, absolving the prisoner in the process.

A mystery remained though: why did blood samples from both brothers match the semen sample, but cheek swabs from only one? Light was shed only with the realization that the prisoner had undergone a *bone marrow transplant* previously. The donor for this transplant was none other than his brother. This made the prisoner's blood genome identical to his brother's; no wonder both matched the sample from the crime scene. But the cheek, and indeed the rest of the body, carried different genomes in the two brothers. The real culprit, the prisoner's brother, could be identified uniquely only by his cheek swab and not by his blood.

In summary, a bone marrow transplant completely replaces the genome in an individual's blood cells. All other cells continue to have the genome the person was born with.

As the name indicates, the bone marrow is a sort of factory located in the interior of our bones. This factory is comprised of so called *stem cells*. Unlike many other cells which can no longer divide, these cells can divide to generate new daughter cells, thus assuring a lifelong supply. Some of these cells then become red blood cells through a controlled process of change. Red blood cells have short lifespans and cannot divide further—meaning that only stem cells in the bone marrow factory can produce new red blood cells. A bone marrow transplant seeks to replace these original stem cells with alternative stem cells from a healthy donor.

Coming back to our story, if *Sita* were to get a bone marrow transplant, the very first step would be to kill her bone marrow stem cells with drugs. Then bone marrow stem cells would be extracted from a donor and transplanted into *Sita*. These cells would divide inside her and some will convert into red blood cells. The hemoglobin in these cells would be manufactured based on the recipe in the donor's genome and not based on the defective recipe in *Sita*'s genome. So, with this one-time procedure, *Sita* would have a lifelong supply of good hemoglobin.

Of course, there are risks with this procedure. Our immune system, which is primarily driven by white blood cells made in the bone marrow, has this uncanny ability to differentiate between our own cells, and foreign cells that come from another person. Such foreign cells are typically attacked and destroyed. Accordingly, *Sita*'s immune system could recognize the donor's cells as foreign and could proceed to attack and destroy them. To prevent this situation, *Sita*'s immune system would be suppressed by killing the stem cells in her bone marrow before the transplant.

The converse issue could be more problematic though. White blood cells created by the donor's stem cells might turn upon *Sita*'s cells, mistaking them to be foreign.

Why would they do so? It turns out that a collection of genes called the *HLA* genes plays a key role here. Every individual has several variants in these genes. If these variants in the donor differ from those in *Sita*, the chances of the donor's cells turning upon *Sita*'s cells increases. So, it is important for donors to be HLA-matched to the recipient. Indeed, a bone-marrow transplant offers *Sita* a greater than 80% chance of complete cure provided the donor is related and HLA-matched with *Sita*.[81] If a related, HLA-matched donor is not available, then an unrelated, HLA-matched donor might still provide a cure, but with slightly lower odds of just above 70%.[82] In either case, there would be no further need for blood transfusions and no progressive complications.

Unfortunately, even with related, HLA-matched donors, there is at least a 6% risk of things going horribly wrong, leading to death.[83] Does *Sita* have any other options?

Editing the Blood Genome

Wouldn't it be wonderful if we could somehow edit *Sita*'s blood genome and restore the problematic G in her *HBB* gene back to a T? We would no longer need to kill *Sita*'s bone marrow stem cells and replace them with someone else's stem cells. We would simply extract some bone marrow stem cells from *Sita*, kill her remaining bone marrow cells, edit the genomes in the extracted cells, and inject them back into her.

Imagine for a moment that we can do so. Then several of *Sita*'s bone marrow stem cells will now have a T instead of the G she was born with. These cells will divide to generate many more daughter cells, each of which will also inherit the T. As some of these cells differentiate into red blood cells, this T will ensure that the *HBB* recipe produces healthy hemoglobin, capable of delivering the required oxygen to her body. And since the rest of her genome remains unchanged, *Sita* wouldn't need to contend with any immune system mismatches. The cure would be permanent and her thalassemia would be behind her forever.

This is easier said than done though. Indeed, how do we take a few cells and edit one chosen character among the billions of genomic characters in these cells? Editing files on a computer is easy. Editing a printed book, less so. Editing a real, physical molecule inside a live cell? Even less so.

A major hurdle in editing the genome inside a live cell is to locate the point where the edit needs to be made—specifically, the position which carries the T→G variant. The editing mechanism has to locate this position from among three billion or so genomic characters. This is a tall order by any stretch. Fortunately, nature has provided us with tools to perform this search quickly and accurately. Interestingly, these tools came to our notice from rather unexpected quarters.

In the last decade or so, scientists have observed an interesting form of immunity in the genomes of bacteria attacked by a virus. The attacked bacterium has its own distinctive genome sequence, as does the attacking virus. Surprisingly, after the attack, stretches of characters from the viral genome were found inserted into the bacterial genome. The bacterium had thus managed to save the identity of the virus into its own genome. Going further, it even used this saved identity to ward off similar virus attacks in the future.[84,85] Here is how.

A special gene called *CAS9* is instrumental in this process. Proteins derived from the *CAS9* recipe can take these saved viral stretches, match them against the genomes of invading viruses, and clip these viral genomes at precise positions where matches are found. So, when the same virus attacks again, *CAS9* invariably matches and clips the virus genome, thus destroying the virus and protecting the bacterium—a very elegant form of immunity that learns as it observes.

Note this remarkable feature of *CAS9*—if you give it any sequence of roughly 20 genomic characters, then it will fish out matching positions in the entire genome for you. Not just in bacterial genomes, but in human genomes as well. The human genome is huge, three billion characters long, but that doesn't discourage *CAS9*. It finds the matching positions and then clips the genome at these positions.

How can we use *CAS9* to edit *Sita*'s genome? We know her genome sequence, so we can carefully design a sequence of about 20 characters and hand it to *CAS9* so that it clips her genome just near the position that needs to be edited—the position of the T→G variant. In fact, using *CAS9* with two distinct sequences of 20 characters each,[86] one can get two clips, one on either side of this position, as shown in this picture.

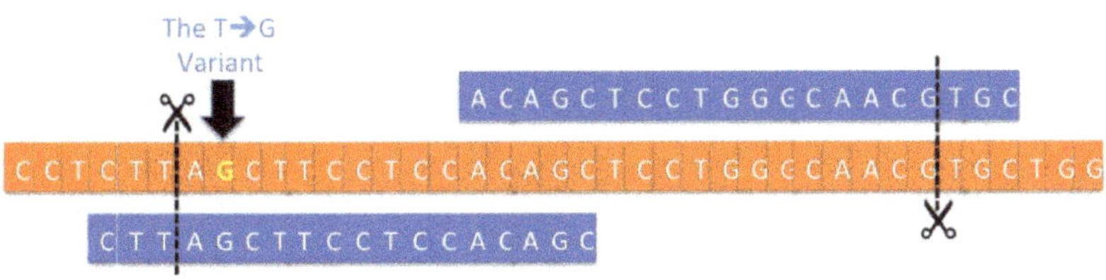

CAS9 locates matches for the two sequences and clips the genome three characters into each sequence.

What next? Genome clipping is actually a common event that occurs routinely in nature. To protect the genome from the impact of this clipping, nature has provided cells with repair mechanisms that help rejoin the clipped portions. We will see glimpses of these mechanisms in a later story. One of these repair mechanisms can be fooled into replacing the genomic stretch of characters between the two clipping points with a new chosen set of characters. By carefully choosing these characters, the aberrant G can be edited back to T.

Since the realization that genomes can be edited using this so-called *CRISPR-CAS9* method,[85,87] much has been done to hone it further, and much effort is ongoing to make it more robust. There are several promising signs already. For instance, scientists have successfully edited genomes in the liver cells of live, adult mice, curing them of a disease called tyrosinemia in the process.[88]

Of course, challenges remain. There is some danger that *CAS9* might clip the genome at unintended places, with unexpected consequences. Or that the repair mechanisms that rejoin clipped portions might not do their job quite as expected. Unfortunately, this approach is not yet ready for medical use. Yet, there is every hope that further improvements will allow us to edit genomes of patients like *Sita* sooner rather than later.

Wrapping Up

The vast expanse of the genome is sparsely dotted by exons carrying gene recipes. These recipes account for a paltry 1–2% of the entire genome. The remainder of the genome comprises introns and regions between genes, neither of which carry recipes. It would appear, therefore, that genomic variants in these regions are inconsequential. But *Sita*'s story tells us otherwise.

Genomic characters in the introns guide the recipe execution process as it jumps from one exon to the next. Variants here could well confuse these jumps. And a jump gone awry could well yield a dramatically different recipe, as seems to be the case with *Sita*.

Indeed, the *HBB* gene is notorious for jumps gone awry on

account of such intronic characters. For instance, 20% of the beta-thalassemia cases in China are caused by a variant sitting deep inside the second intron of *HBB*.[89] This variant causes the splicing jump to land prematurely in the middle of this intron; recipe execution resumes from this point and continues for a while before jumping again to the end of the intron, where it should have landed in the first place. Of course, the recipe is modified dramatically as a result.

Interestingly, modified recipes in *HBB* only manifest themselves several months after birth. Until then, a form of hemoglobin called *fetal hemoglobin*, made by combining recipes in the *HBA* and the *HBG* genes, holds the fort. Six months or so after birth, fetal hemoglobin is phased out and regular hemoglobin kicks in. It is then that the ill-effects of a modified *HBB* recipe come to the fore.

Meanwhile, *Sita*, who has had numerous transfusions, awaits a suitable donor for a bone marrow transplant. A suitable donor presents a good chance of successfully correcting *Sita*'s thalassemia without introducing any other complications. Even so, there is always a small risk of things going wrong.

Hopefully, the ability to edit genomes will become a reality in the not too distant future, removing the need for donors and associated complications. *Sita*'s doctor did not need to know which variant *Sita* had in her *HBB* gene to diagnose her thalassemia. But precise knowledge of this variant would be needed for editing her genome. And that knowledge is now available.

Story 7

The Ominous Reflection

In this story, we shift gears to cancer, a scourge we are all familiar with. The subject of our story is two-year-old *Ali*, who was bothered by redness and irritation in his eyes. His parents noticed a squint along with an odd whiteness in the center of his iris. They consulted with a doctor who examined the insides of his eyes.

The doctor first dilated *Ali*'s pupils (those little holes through which light enters his eyes) and then shone light through them. He expected this light to reflect back from *Ali*'s retina and appear as a reddish-orange circle around a tiny white dot. Here is what he saw instead.

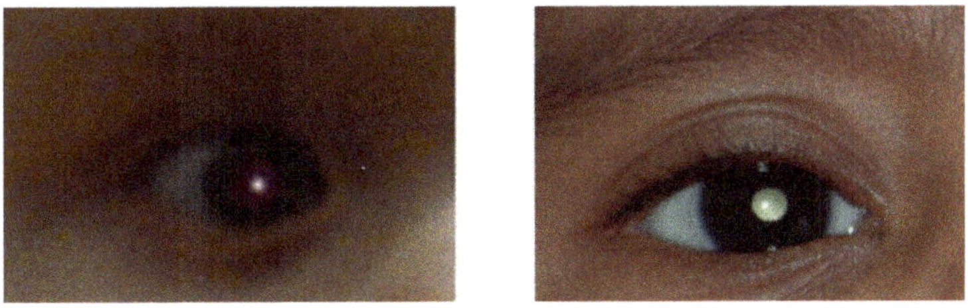

The expected reddish-orange reflection on the left,[90] *and the reflection from Ali's eyes, on the right.*[91]

The large white spot instead of the reddish-orange one was unexpected. Surely, there was a problem in *Ali*'s retina. Leaky blood vessels were a possible cause; a serious one at that, for they could lead to blindness. But there was another, far more serious, possibility. Further investigation revealed that the large white spot came from a tumor in *Ali*'s eye, a potentially life-threatening cancerous growth, called *retinoblastoma*.

We usually think of cancer as a disease of aging. It appears most often in one's 60s and 70s. Cancer in infancy is rare. Retinoblastoma, even more so; only around 350 new cases are diagnosed each year in the United States, and around 1500 in India. *Ali* was one of these unfortunate few. Some cells in his retina had turned rogue, defying the social contract that binds them in discipline with their neighboring cells. These bad citizens were dividing uncontrollably and destroying the organized cellular arrangements needed for proper vision. If left untreated, blindness was imminent. But that was not all.

Cancerous cells often leave the organs they belong to and wander off to other parts of the body, carrying their destructive effect with them. When this happens to organs vital for life, say the brain or the bone marrow, life itself gets threatened. Indeed, the large white spot was a threat to *Ali*'s life itself. What caused this life-threatening condition in *Ali*, and that too at such a young age?

Heeding the Warning

The white spot, though ominous, can be a life-saver by providing early warning. A leading newspaper in Bangalore carried the following heart-warming story on May 12th, 2012,[92] as related to us by Ashwin Mallipatna, a specialist in retinoblastoma, practising at Narayana Nethralaya, Bangalore.

A couple with a young daughter in the suburbs of Bangalore noticed the white spot in their daughter's eye when she was 6 months old. The spot was still very small, so no alarm bells were rung; on the contrary, it was interpreted as a good omen.

A few months later, they took their daughter to a photo studio for her to be photographed for an occasion. Typically, the pupil of the eye shows as a red dot when the face is photographed. This happens on account of reflection from the blood vessels in the retina. Modern cameras have sophisticated methods to correct for this redness so you may not always spot it on a photograph. However, when a retinoblastoma grows on the retina, the reflection no longer appears as a red dot; in its place instead is the dreaded white spot, much more prominent on a photograph than it is to the naked eye. Indeed, the daughter's photograph showed a very prominent white spot.

Fortunately for the family, the photographer did not think this white spot was a good sign. He was well aware of its ominous nature, having lost his two-year-old niece a few years ago to retinoblastoma. Accordingly, he suggested that the daughter be shown to a specialist. Just in time, it turned out, for appropriate treatment saved her eyes and her life.

Was *Ali* so lucky as well? We'll see. But a question first: what causes tumors in the eyes of *Ali* and other little infants like him?

Inheritance or Bad Luck?

Was *Ali*'s retinoblastoma on account of inheritance from his parents? Or was it due to some other factor acquired after he was conceived? Was it a controllable factor or was it just plain bad luck? Some clever deduction from simple patient observations yielded the first insights on these questions in the early seventies.

Doctors had long noticed that some patients with retinoblastoma presented tumors in both eyes (the *bilateral* type), and others in only one eye (the *unilateral* type). These two types appeared to have different behaviors and origins.

The bilateral type appeared clearly hereditary—children of patients who had survived the disease were found to experience a 50% risk of disease themselves. Variants in some yet unknown gene were therefore the likely cause. A parent with the variant in one gene version would pass that defective version on to his or her child with a 50% chance. And children inheriting this defective version would develop retinoblastoma.

However, there was a catch to this explanation. A few infants with bilateral retinoblastoma (1%–10%)[93] exhibited a slightly perplexing trend. The disease did indeed appear to be hereditary in these infants because there was a close relative, a sibling, perhaps, who also had the disease. However, strangely, neither parent showed any sign of the disease. If genomic variants were the cause of retinoblastoma, then these infants would have inherited their problematic variant from one of their parents. Yet, the parents did not develop retinoblastoma. Why?

Maybe an inherited variant in one gene version, while instrumental in causing the disease, was insufficient by itself. There was at least one, and maybe many, additional causes that were needed for retinoblastoma to develop. These causes, whatever they were, seemed to get their way in most but not all individuals born with a problematic variant.

The unilateral cases, with tumors in only one eye, further underscored the presence of these additional causes, for they appeared to not be hereditary at all. Children of parents with unilateral retinoblastoma rarely seemed to develop the disease. Possibly, the additional causes, whatever they were, were flying solo in these cases, and yet strong enough to compensate for the absence of an inherited genomic variant. What were these additional causes?

These observations were made well before sequencing the genome was possible, so scientists had to answer these questions with much simpler forms of data available at the time. To this end, in 1971, Alfred Knudson at the M.D. Anderson hospital in Texas dug deeper into the medical records of 48 patients with retinoblastoma, 23 of whom had bilateral tumors and 25 unilateral.[93] And what did he uncover about the behavior of these unknown causes?

Knudson's Bilateral Cases: The Single-Hit Roulette

Starting with the 23 bilateral cases, Knudson examined the ages at which these patients had been diagnosed. Since diagnosis usually happens within the first two years of life, the age of diagnosis is measured in months rather than years. Take a moment to look at these numbers.

2 3 3 3 4 4 5 6 7 8 9 11 12 12 13 15 18 18 22 24 30 44 60

Do you see a pattern here? If not, then read on. Also note the average age of diagnosis: 14 months.

What do these numbers tell us about the mysterious additional causes of retinoblastoma? Are these causes already present at the time of conception? If so, one would expect tumors to start forming at roughly the same age in all these children. Of course, never exactly the same age, for there is always natural variation—some might start a little earlier, and some a little later. These tumors would then grow at roughly the same rate in all these children until they became discernible, leading eventually to diagnosis at roughly the same age, with some natural variation. Therefore, we would expect the ages of diagnosis to be around the average mark of 14, with some leaders appearing earlier and roughly an equal number of stragglers following later. However, the numbers above show no such pattern.

Instead, our numbers appear tightly clustered at the beginning (many 3s and 4s) and are very sparse toward the end (after 24, they skip all the way to 30, and then even further to 44, and thence all the way to 60). Our assumption that the additional causes were present at the time of conception appears incorrect, or so the numbers say. There must be another explanation.

Maybe the additional causes are acquired after conception, when numerous cell divisions create tens of trillions of cells starting from a single cell. The genomes in each of these daughter cells is certainly a copy of the original, but not an exact copy. Mistakes do happen on occasion, sometimes driven by damage to the genome, as we will see in the next story. Therefore, different cells in our body develop variants which we did not necessarily inherit from our parents. Most of these variants are not passed down to our children (with the possible exceptions of those in our sperm cells and egg cells). These so-called *somatic* or acquired variants live and die with us. Could the additional causes be related to these somatic variants?

More specifically, was the retinoblastoma in Knudson's bilateral cases caused by the combination of an inherited variant and a somatic variant? Of course, all cells in the retina would have the inherited variant. Additionally, one or more such cells could have acquired the somatic variant, well after conception. The presence of both variants in the same cell could then tip the cell over into a cancerous state, leading to retinoblastoma. This is certainly a plausible explanation, provided it matches with Knudson's observations. Does it?

Some Mathematical Jugglery

Answering this question requires that we deal with the randomness that seems to underlie the onset of these somatic variants—one cannot pinpoint with certainty when or why such a variant might arise in any of our cells. Dealing with this uncertainty requires a bit of mathematical jugglery, which you can choose to either skim through or even skip altogether in favor of the next section.

Coming back to randomness—just as the outcome of a die throw cannot be predicted with certainty, the timing of when a person with an inherited variant might acquire a somatic variant cannot be predicted with certainty. However, one can talk about the chances of this happening. This is particularly easy for dice. A single throw of a die has a 1 in 6 chance of yielding a six. So, a die thrown 100 times will yield close to $100/6 = 16.6$ sixes. Similarly, given 100 individuals, each with an inherited variant, we can estimate how many will develop a somatic variant in a given month. This is not so straightforward though, because the odds of a single such individual developing a somatic variant in any given month are not readily known. So, we have to make guesses and eventually determine which guess is most consistent with Knudson's numbers. For purposes of illustration, let us start with a guess of 50%. Then, of the 100 individuals at hand, roughly 50 will develop a problematic somatic mutation in a given month.

What happens to the other 50 who do not develop a somatic mutation this month? How many acquire a problematic somatic mutation in the next month? A moment of thought should tell you that the right answer is roughly 50% of 50, or 25. And roughly 50% of 25, or 12.5, in the third month, and so on. The precise numbers here are not important, for they are based on the somewhat unfounded assumption of 50% chance. What is important, however, is the pattern: a lot initially (50), a little less next (25), even lesser next (12.5), and so on. We can visualize this pattern as a curve that successively flattens out as the months progress.

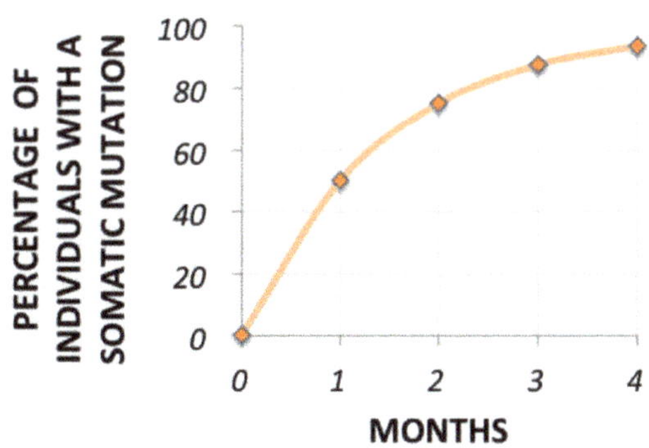

The successively flattening curve depicting the fraction of individuals who acquire a somatic mutation as the months progress.

Interestingly, if the chances of acquiring a somatic variant were not quite 50% each month, the precise numbers would change, but this overall pattern of successive flattening would remain. You can easily verify this by experimenting with values other than 50% for yourself.

In fact, you can go further and determine which value yields a curve that best matches the rate at which Knudson's 23 bilateral patients were diagnosed with disease. If indeed a single somatic variant acting on top of the inherited variant was the cause of the disease in Knudson's bilateral cases, the percentage of his patients diagnosed with time should fall right on this curve. Do they indeed? The picture below gives us the answer.

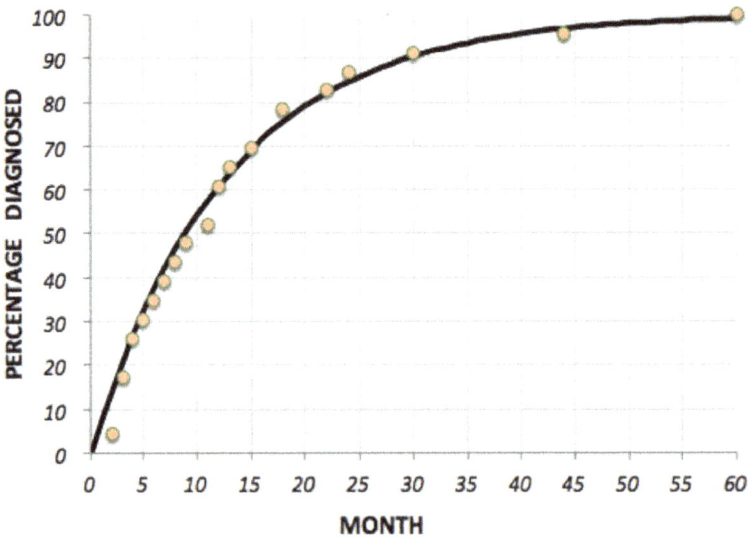

The best somatic-hit curve and the rate of diagnosis for Knudson's bilateral cases.

The dots, indicating the ages of diagnosis of Knudson's 23 bilateral cases, lie smack-bang on the curve calculated using a somatic variant incidence rate of about 7.7% per month. The match between the dots and the curve appears too good to be accidental; there seems to a method to this random madness.

Bad Luck, Almost Always

Indeed, it seems that an inherited variant followed by a single randomly acquired somatic variant was a plausible cause of retinoblastoma in Knudson's bilateral cases. The first hereditary *hit* alone was insufficient; a second somatic *hit* was required. These two *hits*, together, caused retinoblastoma.

Note the element of luck in the second hit. It is not a predictable event—one cannot say with certainty when it might happen and to which cell. Yet, as the curve above suggests, the chance of someone with an inherited variant also experiencing an unfortunate somatic variant in some cell approaches near-certainty: 95% over 36 months, and 99% over 60 months. Therefore, a child inheriting the first hit will almost certainly acquire a second hit as well and then develop retinoblastoma in the first few years of its life. It stands only a slim chance of escaping this second hit in the first 60 months of life.

Of course, given more time, the appearance of a second hit seems inevitable. This suggests that a child with an inherited hit will develop retinoblastoma sooner or later. However, it turns out that children with an inherited first hit who do not develop retinoblastoma in the first few years of life usually escape the disease forever. Something changes in the retina after the first few years. What could this change be?

Do cells in the retina stop dividing after the first few years of life, reducing the chances of accruing somatic variants? Indeed, retinal cells lose the capability to divide after a point. This is good and bad: good because it practically nullifies the possibility of cancer, but bad because cells lost due to injury aren't spontaneously replaced. However, the mystery is that retinal cells actually stop dividing much earlier, well before a person is born. What changes then after the first few years of life, thus nullifying the risk of retinoblastoma? There appears to be more to the two-hit theory than meets the eye.

We postpone this question, as well as our exploration of *Ali*'s first and second hits, until after we have investigated Knudson's unilateral cases as well, for they too have their bit to add to this story.

Knudson's Unilateral Cases: The Two-Hit Roulette

You might recall that unilateral cases appeared not to be hereditary at all. In other words, there was no inherited hit in these cases. Where then did the first hit come from? A moment of thought suggests a possibility—could both hits be random somatic hits? The answer is hidden below in the ages of diagnosis for 25 of Knudson's unilateral patients,[93] if only you can spot it.

5 8 10 15 19 21 22 24 27 28 29 29 31 32 33 34 36 36 38 46 47 48 50 52 73

Note the average age of diagnosis now: 31 months, significantly longer than the 14-month average for bilateral cases. Tumors in unilateral patients seem to be setting in much later.

Not just that, the pattern itself appears to be different. Remember how bilateral cases were—steep at the beginning (many cases diagnosed early) and getting flatter and flatter as time progresses (fewer and fewer cases diagnosed as time elapses). Instead the unilateral cases appear to be slightly flat at the beginning with only three cases in the first 14 months (5, 8, 10). Then it gets a little steep in the middle with as many as six cases in five months (27,28,29,29,31,32). At the end, it again goes flat with just three cases in 23 months (50, 52, 73). Is this the profile one would expect with two random somatic hits?

Note what we mean by two somatic hits—a single cell, among the millions of cells in the retina, has to develop two somatic variants. Two problematic variants in two different cells will tip neither cell over into dividing uncontrollably. Both hits have to be in the same cell.

How fast do individuals develop two random hits in one of their millions of retinal cells? Certainly at a slower rate, for the challenge is harder. In initial attempts, some cells may be hit once each but none would be hit twice. After a while, when sufficiently many cells have been hit once, the chances of hitting an already-hit cell again will become substantial. So, success rates will see slow take-off and then a more rapid rise. This will be followed possibly by the same flattening-off we saw in the bilateral case. This pattern appears broadly consistent with the numbers above. So, we derive the precise curve mathematically and examine how well the rate of diagnosis of Knudson's patients fits this curve.

Unfortunately, this time around, success is more elusive. The curve that results lies well below the actual dots; in other words, Knudson's unilateral cases appear to be diagnosed at a much higher rate in reality than it would suggest. Our hypothesis that two randomly acquired somatic hits cause unilateral retinoblastoma doesn't seem to hold water. How else do we explain these unilateral cases?

More Mathematical Jugglery

This needs some more mathematical jugglery. Again, feel free to skim through what follows in this section or skip it altogether.

So far, it appears that two randomly acquired somatic hits, each occurring with a 7.7% per month chance, do not explain Knudson's unilateral data. We have to try something different. While we do so, we must also preserve the satisfactory explanation that one somatic hit provides for the bilateral case.

To this end, there is one nuance we haven't yet investigated: maybe the chances of the first hit are different from those of the second? Is it possible that once a cell receives its first somatic hit, it becomes more receptive, or possibly less receptive, to the next somatic hit?

Accordingly, we experiment with different odds for the first somatic hit, while keeping the second hit consistent with the bilateral case. At a success rate about 40 times that for the second somatic hit, our curve starts to match Knudson's numbers well. That's progress. Unfortunately, the match lasts only for the first three years; thereafter, our curve shoots way above the dots. Among Knudson's patients, there are *many fewer* cases diagnosed after three years than our hypothesis suggests. We are still missing something.

Remember, we observed earlier in the bilateral case that something changes in the retina after the first few years. The same phenomenon appears to be resurfacing here. The chances of a second somatic hit appear to be weakening after the first three years. So, we recalculate our curve allowing odds for the second somatic hit to taper off gradually after three years, and reducing to negligible levels after five years. How well does this recalculated curve fit the dots? Let's take a look below.

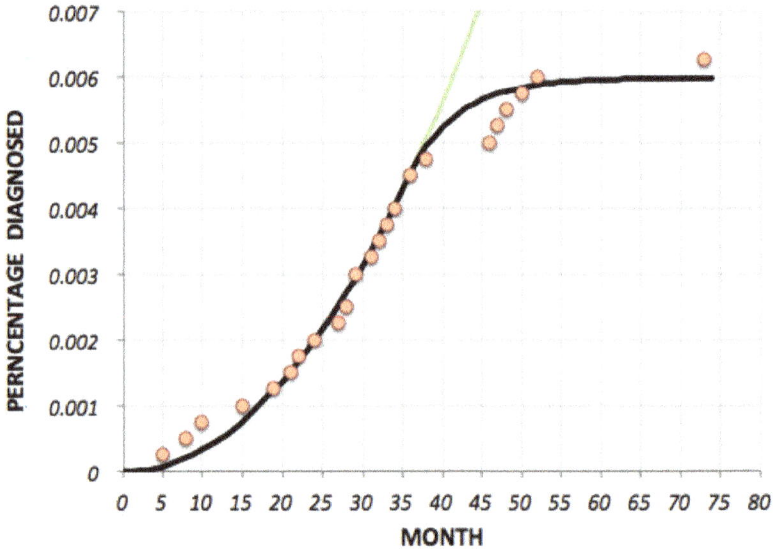

The two-hit curve and the rate of diagnosis for Knudson's unilateral cases. The light curve continuing upward indicates what happens if the odds of a second somatic hit do not fade out after three years.

The dots, indicating the ages of diagnosis of Knudson's 25 unilateral patients, match the two-hit curve fairly well now. This is reconfirmation of the suggestion that two randomly acquired somatic variant hits with differing odds together form a likely cause for unilateral retinoblastoma.

The Method behind the Random Madness

To summarize, the analysis of Knudson's bilateral cases suggests that those who inherit a problematic variant invariably experience a second random somatic variant in one of their millions of retinal cells in the first few years of life. So high is the likelihood of this second hit that multiple such hits often occur in different cells, leading to bilateral cancer. Fortuitously, a small fraction of individuals scrape through the first few years of life without any second hit at all; they then escape the disease altogether.

Of course, most people are born without any inherited hit whatsoever. The analysis of Knudson's unilateral cases suggests that one in 16,000 or so individuals will be unfortunate enough to receive two somatic variant hits in the same retinal cell, simply by sheer chance, leading to unilateral cancer.

Isn't it amazing that the analysis of just a handful of observations can yield so much insight? Such is the power of good data. This data also raises further questions. The two hits seem to have different likelihoods of occurrence—the first appears 40 times more likely than the second. Why so? What makes the second hit different from the first? And why do the chances of the second hit diminish from modest to negligible between three and five years of age?

We will get back to these questions shortly. But first, back to *Ali*'s case. His retinoblastoma was bilateral, which made it likely that he had inherited a problematic variant from his parents. However, neither of his parents had retinoblastoma. What could explain this mystery? Further, *Ali* had suffered not one but at least two somatic hits, one in each eye. Which gene did these hits occur in? And why did these variants send his retinal cells into a flurry of uncontrolled division?

We have several questions to answer. The first stop in our quest for answers is the core issue in cancer—the division cycle of a cell.

The Division Cycle of a Cell

The act of cell division has been perfected by nature over a billion years—well, we should be careful and say nearly perfected, for *Ali*'s travails, and cancer in general, do indeed stem from errors in this cell division process.

Before a cell divides into two, there is much to be done in preparation. First, the cell grows larger and accumulates enough material for both its daughter cells (call this the *first growth phase*). Once there is enough, the cell makes a complete copy of its entire genome (the *copy phase*). It then grows a bit more (the *second growth phase*). And then comes the time for the cell to split itself and all its assets, including its duplicated genome, among its two daughter cells (the *split phase*). These daughter cells then undergo the same cycle, in due course.

The transition from one phase to another in this cycle requires different sets of genes to be activated. For example, take the transition from the first growth phase to the copy phase. Hitherto dormant genes capable of copying the genome must be woken up at this point and their recipes executed. To this end, nature has evolved an elegant switching mechanism: proteins derived from a family of genes called *E2F* attach themselves to specific sites in the genome, triggering recipe execution on genes located close by; these happen to be genes instrumental in the copy phase.

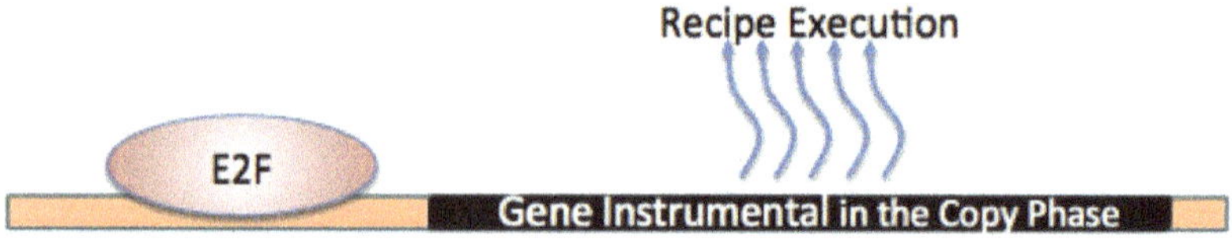

Nature's switch: E2F proteins attach to the genome at specific positions, and initiate recipe execution on genes instrumental in the copy phase.

And this leads to the next question—what keeps the *E2F* proteins in check during the first growth phase, so they do not usher in the copy phase prematurely?

This task is left to another key molecule, which takes center stage in this story. This molecule tethers itself to the *E2F* proteins, thus restraining them from attaching to the genome. This holding pattern continues until the cell has grown sufficiently. When sensors in the cell detect that sufficient growth has happened, for instance, when the cell starts touching its neighboring cells, our key molecule lets go of the *E2F* proteins, allowing them to attach to the genome and initiate the copy phase.

How does our key molecule know when to let go of the *E2F* proteins? This is yet another of nature's very elegant switching mechanisms. Our key molecule is a protein comprising many amino acids, one of which is typically embellished with a special chemical mark—a negatively charged phosphoryl group (PO_3^{2-}). In this singly-marked form, it tethers itself to the *E2F* proteins,[94] as in the picture below. When the growth sensor detects that sufficient growth has occurred, a cascade of events leads to marking of many more amino acids in this protein. This forces a change in its electrochemical properties and thereby a change in its shape, forcing the *E2F* proteins loose in the process.

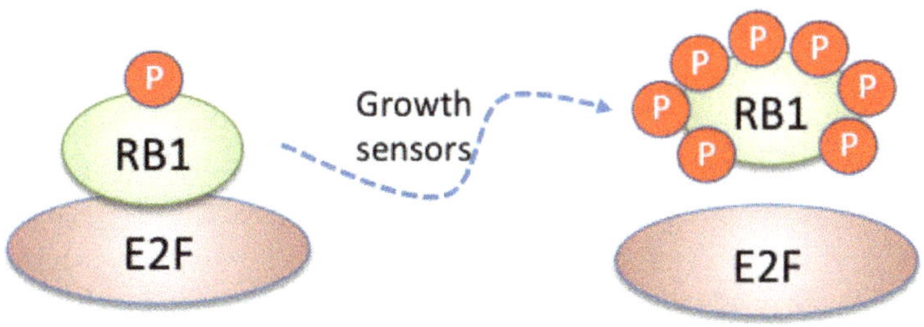

Another switch: our key molecule lets go of E2F when it gets many marks.

Which is the key molecule referred to above? This is the protein obtained from a gene often found to be aberrant in retinoblastoma patients. This gene has now been confirmed as the causative gene in retinoblastoma. Not surprisingly, it is named *RB1*.

Could two hits in *RB1*, one inherited and another somatic, have incapacitated the gene in *Ali*? If so, then the *E2F* proteins would be free to rush cells into the copy phase prematurely, thus accelerating cell division and causing tumors as a consequence.

A Peek at *RB1*

To answer this question, we need to peek into *Ali*'s genome and identify any hits he might have on the *RB1* gene. We can do so by sequencing his genome, as we did in our previous stories. In most of these stories, *Sita*'s thalassemia story being an exception, we cast our net wide and sequenced all exons from all 20,000 odd genes, or even the entire genome. Casting such a wide net in *Ali*'s case is overkill, for his problems are very likely to stem specifically from the *RB1* gene.

However, unlike the *HBB* gene in *Sita*'s case, which was all of 1605 characters long, the *RB1* gene is huge: approximately 200,000 characters long. It comprises 27 exons, which carry the gene recipe, along with 26 intervening introns. Sequencing such a large gene in its entirety is cumbersome and expensive. Fortunately, its 27 exons together add up to just 2783 characters; the remaining characters are in the introns. Since the recipe of the gene is *focussed* in these 2783 exonic characters, why not sequence these alone? Indeed, that is what we do, and it is far less expensive. Not that it matters for this story, but we actually spread the net a little wider to include all exons from roughly a hundred genes that cause hereditary cancers.

So, we extract *Ali*'s genome from a few drops of his blood and sequence exons from the selected genes. We then turn eagerly to *RB1*, looking for problematic variants in this gene.

Remember the different types of variants we encountered in previous stories. There were missense variants, where the usual amino acid in the gene recipe is replaced by another. Then there were nonsense variants, where the recipe change is even more dramatic; the recipe stops dead in its tracks prematurely. There were also frameshift variants, where the grouping of the recipe into triplets gets jittered due to extra or missing characters, whose count is not a multiple of three. Unfortunately, we find no such variants of note in *Ali*.

How about variants which lead the recipe execution process astray as it makes splicing jumps from one exon to the next? We saw a variant of this type in the story on *Sita*'s thalassemia. Of course, that variant was in an intron, and not in an exon. And we haven't particularly focussed on introns for our sequencing. Fortunately, we have actually sequenced a bit more than just the exons of *RB1* in *Ali*, for each exon we sequence drags along 50–100 characters of the neighboring introns on both sides. Frustratingly though, our exploration of these intronic portions yields no suspects either. Our variant continues to elude us.

It is possible that our variant hides deeper inside the intronic ocean, far away from the shores of any of the 27 exons of *RB1*. Our sequencing hasn't plumbed these depths, for dredging this intronic ocean is considerably more expensive. That is a less likely scenario though. Is there a more likely scenario that we are missing?

Missing Chunks, but with a Difference

Recall the case of *Raj* and *Taj* and the odd placement of their internal organs. The cause of this phenomenon was a large missing chunk in their genomes. Along similar lines, could there be a large chunk missing from the *RB1* gene in *Ali*? Somewhat along the lines shown below?

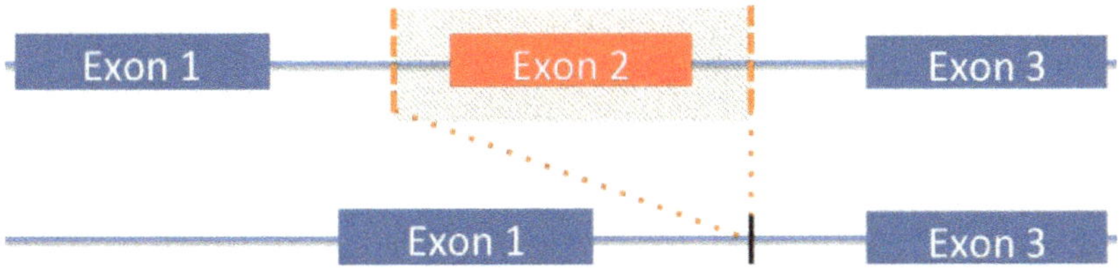

A hypothetical missing chunk in Ali, below, as compared to the reference sequence, above. The missing chunk comprises exon 2 along with some sandwiching intronic portions.

How would we answer this question? Let us turn back to how we found the missing chunk in the genomes of *Raj* and *Taj*. We took each read in turn and searched for it in the reference sequence. This search was successful in many cases but did fail for a few reads. Hidden amongst these were reads that straddled the missing chunk. We split such reads into two parts and searched for each part separately in the reference sequence.

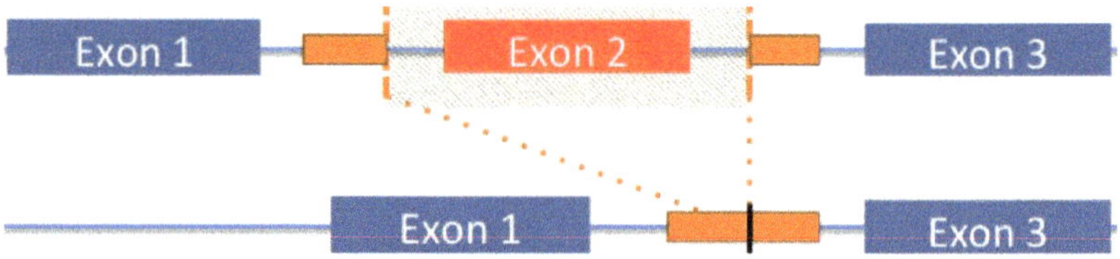

Reads straddling the hypothetical missing chunk in Ali's genome, below, are split when matched to the reference sequence at the top.

If *Ali*'s genome was indeed missing exon 2, then these two parts, though adjacent in *Ali*'s genome, would find their respective matches far apart on either side of exon 2, as in the picture above. And this would tell us that the stretch in between was missing from *Ali*'s genome. With much hope, we turn to this strategy on reads from *Ali*'s genome. This time around, our success in *Raj* and *Taj*'s case doesn't repeat. A moment of thought then exposes a catch, a rather serious one, unfortunately.

Unlike for *Raj* and *Taj*, where we sequenced the whole genome, we have sequenced only the exons of *RB1* in *Ali*, i.e., exons 1 and 3. Additionally, we have nibbled a bit into the introns. Only a bit, and not too much. So, we have no reads at all in *Ali* which straddle this hypothetical missing chunk that appears in the middle of an intron. And without any such reads, this method is a non-starter.

Is there a way to determine if the first hit on the *RB1* gene in *Ali* was due to a missing chunk, without having any reads at hand that straddle the missing chunk, and without having to spend much more to sequence the whole genome?

The First Hit

Fortunately, there is a way. *Ali* has two versions of the *RB1* gene. When we sequence all exons of *RB1* from his genome, both versions contribute reads. Typically, they do so in roughly equal measure, for there is no reason, other than natural variation, for one version to contribute more than the other. However, if one of these two versions has a missing chunk, then the picture changes. As you might guess, the version with the missing chunk will contribute fewer reads. Does this yield a clue?

Let us see. Suppose *Ali* had a missing chunk that took, say, exon 25 away from one of his two versions of *RB1*. If we expected 500 reads to match at exon 25 typically, we would expect to see only 250 now. Or roughly thereabouts, for these numbers do have natural variation.

Accordingly, we take a quick look at the counts of reads placed against each of the 27 exons, with the hope that one of these counts will be much lower than the others, thereby confirming its presence in the missing chunk. What does this picture tell us?

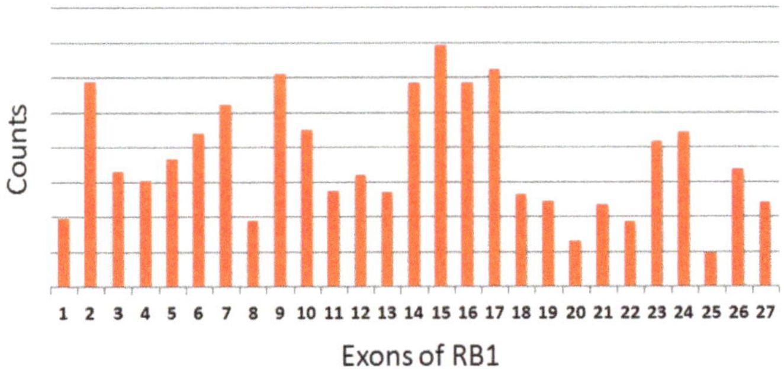

Counts of reads matching to various exons of RB1 in Ali, adjusting for the fact that exons have different lengths and might yield different read counts due to this alone.

Well, exon 25 has a low count. This suggests that it may be

missing from one version of *Ali*'s genome. But then, exon 20 too has a low count. Maybe both exons are part of a larger missing chunk that encompasses all exons in between? Unfortunately, throwing water on this hypothesis are exons 23 and 24 with high counts. The missing chunk, if there is one, is certainly not announcing itself very clearly in this picture. Maybe there is no missing chunk after all.

Before we give up, it is worth one last look. This time, we compare counts of reads matching the various exons in *Ali* with the corresponding counts in several other normal individuals whose genomes we have previously sequenced. This is a more apple to apple comparison—exon 25 in *Ali* compared with exon 25 in several others, rather than exon 25 in *Ali* with other exons in *Ali*. And what do we find?

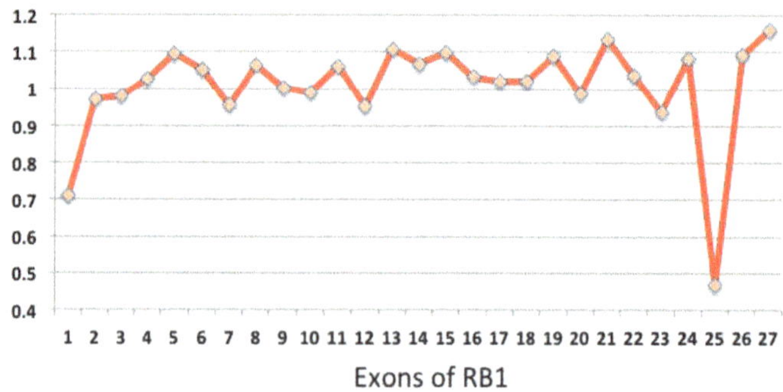

Ratios of Ali's counts relative to the average person, for each exon.

Here, staring at us, almost mockingly, is the first hit on the *RB1* gene in *Ali*: exon number 25, dipping down to the 0.5 mark, leaving behind all the other exons around the 1 mark. *Ali*'s read count is half that of most others for exon 25. Surely, there is a missing chunk around exon 25 in one of *Ali*'s two *RB1* gene versions.

The boundaries of this missing chunk are probably deep inside the introns. We can't get to these boundaries because we have not sequenced so deep into the introns. But we know for sure that exon

25 is part of this missing chunk. This was the first hit on the *RB1* gene that *Ali* was born with. Which was the second?

The Second Hit: *Crossing-over* Again

Remember, *Ali* had bilateral tumors. So, at least one cell in the retina of each eye had incurred a second hit in the *RB1* gene. As Knudson's data indicated, these second hits in *Ali* were not inherited; rather they were introduced after conception by a random process. What were these hits?

We face a challenge in answering this question. Only the first inherited hit is visible in the genome extracted from the cells in *Ali*'s blood. This hit is present in every cell in *Ali*'s body, so it is present in the cells in his blood as well. In contrast, the second hits are present only in a few cells in *Ali*'s retina. These cells, possibly along with others, comprise the tumors in *Ali*'s eyes. To identify these second hits, we will need to extract the genome from *Ali*'s tumor cells and then sequence this genome.

Extracting tumor cells from *Ali*'s tumors requires surgery, which is far more complicated than just drawing a few drops of blood. Surgery of this type to identify variants in the tumor genome is an important element of our endeavor and will be the subject of a later story. For *Ali*, we will have to guess what happened without necessarily peeking into his tumor genome.

Of course, we guess that this second hit must have compromised the other version of *RB1* in *Ali*'s tumor cells (the version not stricken with the first hit). With about a 50% chance, this would have happened in one of two ways.[95] Possibly, the second hit could have removed this other version of *RB1* altogether. Alternatively, it could have mimicked the first hit on to this other version.

"Really!", you might ask. What are the odds of that? How does the second hit manage to mimic the first one?

The answer lies, possibly, in *crossing-over*—the same phenomenon we encountered when we explored my color blindness in the very first story. A quick review might help you recall what this was.

We have two versions of each of our chromosomes, keeping aside the special X and Y chromosomes. Which of these two versions does a parent pass down to his or her child? The answer, if you recall, is neither. What is passed down is a hybrid of the two versions. This hybrid is created when germ cells in our testes or ovaries divide to create sperm and egg, respectively, by a process called *meiosis*.

Cells in our retina and other parts of the body divide using a different process, called *mitosis*. The goal of this process is to create new cells within an individual, rather than create cells capable of developing into a whole new individual. A similar hybrid formation as in meiosis occurs in mitosis as well, albeit much more infrequently—in fact, about 100 times more infrequently.[96] Yet not infrequently enough to spare individuals like *Ali* who already had a first hit. Here is what might have happened to *Ali*.

Imagine one of the cells in *Ali*'s retina as it goes through mitosis. Now focus on chromosome 13, the chromosome that contains the *RB1* gene. This cell has two versions of chromosome 13, one inherited from each of *Ali*'s parents. Accordingly, we call these versions *P* and *M*—short for paternal and maternal, respectively. The character sequences in these two versions are slightly different. Indeed, only one of these, say *P*, carries the first hit; the *M* version doesn't. The picture below shows what happens when the genome is copied now.

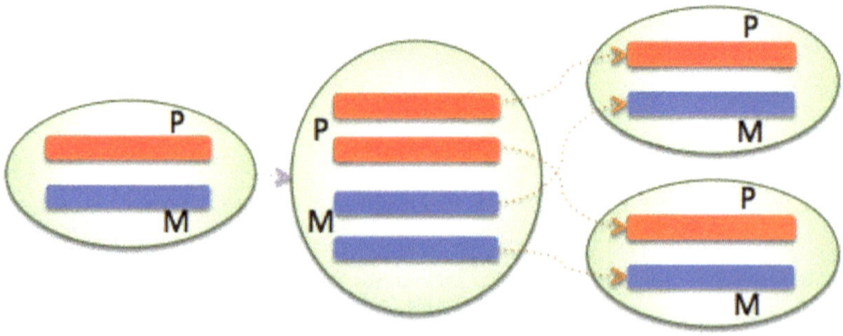

Genome replication when a cell divides.

As depicted above, a copy is created of each version, and then these copies are split among the two daughter cells. Usually, each daughter cell gets one of the two *P*'s and one of the two *M*'s. The good *M* copy protects each of the daughter cells from the inadequacies of the first-hit stricken *P* copy. So, usually, all is well in both daughter cells, in spite of the first hit.

Occasionally, a freak crossing-over event spoils this harmony. This freak event usually occurs before the cell splits into two.[97] One of the *P* copies exchanges segments with one of the *M* copies, as shown below. This *M* copy now carries a segment of *P*. The *P* copy, in turn, carries a segment of *M*.

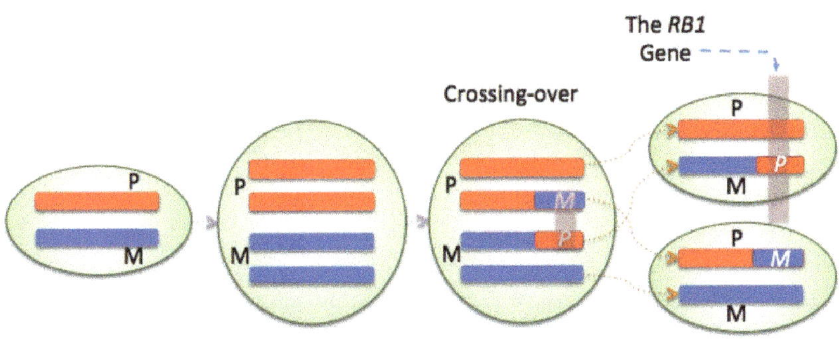

Crossing-over leads to a second hit in RB1 that mimics the first.

The stage is now set for the problem to unfold. As the picture above shows, one daughter cell gets a copy of *P* along with the impure copy of *M*—the one with a bit of *P* in it. If the *RB1* gene happens to be located within this exchanged segment, the consequence is most unfortunate—this daughter cell now has the first hit in both versions of the *RB1* gene. Thus, we have a second hit in this cell that mimics the first.

Some cells in *Ali*'s retina might have lost exon 25 from both their *RB1* gene versions, possibly due to such an event—an event rarer than one in a million maybe. But with several million cells in the retina, what are the odds of this? Not that low. In fact, high enough for this to happen in cells from both eyes.

Then what? How did the loss of exon 25 in both versions of the *RB1* gene in these cells lead to *Ali*'s retinoblastoma?

Password to Nucleus Lost?

The *RB1* gene has 27 exons. The last two exons are rather short, so exon 25 is very close to the end. What happens to the *RB1* gene recipe when this exon is no longer present?

The answer lies in the triplets comprising this recipe. It so turns out that the last triplet in exon 25 overflows into exon 26. And that spells trouble when exon 25 is lost, as shown below.

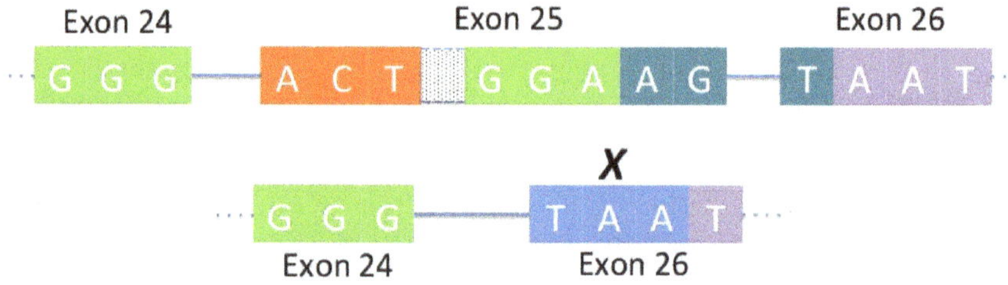

The RB1 recipe, with exon 25, above, and without exon 25, below. The triplets in exon 26 change dramatically.

Without exon 25 in *Ali*, the triplets in exon 26 and beyond are jittered, leading to a completely new sequence of amino acids. And if that's not drastic enough, the first new triplet in exon 26 ends up as a *Stop* triplet, signaling the end of the recipe. The *RB1* gene recipe thus comes to a grinding halt after exon 24.

Remember, in *Tara*'s story of heart failure, we encountered a mechanism called *nonsense-mediated decay* which cells use to destroy proteins created from such prematurely truncated recipes. If this mechanism were to apply, then the first and second hits together would leave *Ali* with no *RB1* proteins at all. Without these proteins, dividing cells will rush through their division cycle much faster than they should.

The picture is complicated though by the fact that exon 26 is the penultimate exon, and a short one at that, at just 50 characters. Premature truncations within 50 characters of the end of the penultimate exon sometimes do escape destruction. So, it is possible that *Ali* continued to have the normal amount of *RB1* protein, just that these proteins were all truncated at the end of exon 24. Were these truncated proteins crippled enough so they could no longer check cell division, thus causing tumors in *Ali*?

Our exploration of this question leads to some interesting answers. The genome, being the precious treasure-chest of gene recipes, requires careful protection inside the cell. This protection is provided by a spherical compartment in the cell we know as the *nucleus*. The nucleus is covered by a membrane that ensures that the genome stays inside and that other molecules outside do not easily get in. And this poses a challenge to the truncated *RB1*.

To appreciate the challenge, let us take a closer look at how the *RB1* recipe is executed. Of course, this action begins inside the nucleus, as depicted in this picture.

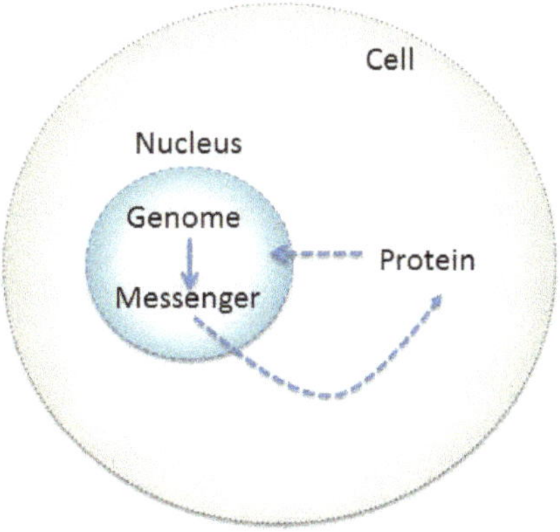

Recipe execution: out and back into the nucleus through pores.

First, the exons and introns of *RB1* are read and copied into a new molecule. Splicing jumps then remove the introns, leaving only the exons concatenated to each other. This modified molecule is called the *messenger* because it carries the recipe from within the nucleus to the outside.

Outside the nucleus, this messenger is read again, and successive triplets of characters are converted to their corresponding amino acids. The resulting chain of amino acids then folds into a characteristic shape to yield the *RB1* protein. This protein must now find its way back into the nucleus as shown in the picture above, so it can rein in the *E2F* molecules from attaching to the genome and ushering in the copy phase of the cell cycle prematurely.[98]

Blocking its march is the membrane of the nucleus, whose explicit goal is to stop gatecrashers. Little pores in this membrane do allow for some traffic; however, they are not large enough for the *RB1* protein to get through. How does it get through then?

For this, it turns to other proteins, known as helper proteins, for assistance. These helper proteins force their cargo through the pores in the membrane of the nucleus. However, they are very selective, so not all cargo is equally eligible. The cargo must carry a password, a specific sequence of amino acids, for it to be eligible to pass. Helper proteins recognize this amino acid sequence and then aid in combining the cargo with other proteins that are then jointly ushered into the nucleus. Typically, the *RB1* gene recipe includes this password: two sequences of characters located right in the middle of exon 25.[99]

With exon 25 lost in *Ali*, this password to the nucleus is lost. The *RB1* protein now finds itself stranded at the doorstep of the nucleus. Some of it does manage to sneak in quietly. But overall, there is lesser *RB1* protein in the nucleus than outside,[99] which is the opposite of what is typical. This inverted distribution of *RB1* has indeed been noted several times in cells that have turned cancerous.[98] No wonder, the two hits which take away exon 25 from both gene versions in *Ali* set the stage for his cancer.

What happens next? Does a cell which experiences these two

hits turn cancerous? Or does something else stand in its way?

Programmed Death

At the heart of this question lies a mystery. Children born with an inherited hit in *RB1* have a greater than 90% chance of developing retinoblastoma in the first few years of life. However, their chances of developing any other type of cancer are much lower, just 48% by age 50.[100] This is unexpected, for the *RB1* gene is known to stem the mad rush to divide in most cells in our body. Yet two hits in these other cells seem less likely to cause cancer than two hits in a retinal cell. Furthermore, two hits in a retinal cell also cause cancer only in the first few years of life. Why?

We know only bits and pieces of the answer today. In essence, nature poses further hurdles that prevent such impatient cells from marching on in their quest for unhindered division. These hurdles vary from one cell type to another.

One such hurdle is best exemplified by comparing a duck's paddle-feet to our fingers and toes. You will surely notice the key difference: webbing. Ducks' feet have digits which are connected together. Human fingers and toes lack this webbing and can therefore move independently of each other.

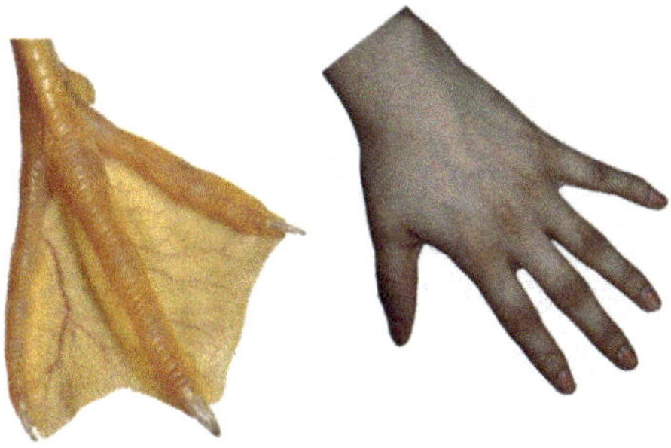

Webbing in ducks, no webbing in humans.

We may never realize this, but our fingers and toes do indeed start off with this webbing. Somewhere along the way, well before birth, this webbing is lost. Six to eight weeks from conception, when we are still tiny foetuses, the cells which form this webbing just die, leaving only the cells forming fingers and toes to continue into full-sized digits.

How do cells in the webbing know they have to die? And why do they do so voluntarily, while their neighbors continue to live and thrive? Alan Turing,[101] the famous mathematician and computer scientist, postulated a mathematical model that provides a possible explanation. According to his theory, all cells together secrete substances which then spread out over the entire area spanned by the digits and the webbing. Interactions between these substances cause their concentrations to become highly non-uniform—some regions build up higher concentrations and others find themselves with lower concentrations. Cells in regions with, say, higher concentrations, then switch on recipe execution on specific genes that cause these cells to die.

This so-called *programmed death*, the unselfish ability of cells to undergo organized suicide in response to various triggers, is nature's way of counterbalancing excessive cell division. And nature maintains this balance using several double-edged swords. The protagonist of our plot, the *RB1* gene, is one such.

Just as the *RB1* protein puts a brake on cell division, it also puts a brake on programmed death. With its two hits in *Ali* reducing the amount of *RB1* in the nucleus, the brakes on cell division are indeed removed. But so are the brakes on programmed death. So, *Ali*'s two hits do not lead to cancer right away. Instead, cells with these two hits typically die.[102]

And that takes our mystery one step deeper, for *Ali* definitely had tumors. Not just one, but two, with one in each eye. How did these come about?

Divide, Die, Differentiate?

The answer appears to lie in *differentiation*, the third piece to our cellular puzzle—the first two being division and death. Our cells aren't all the same; rather, they are differentiated in that different cells specialize in different functions. Heart muscle cells specialize in contraction, retinal photoreceptor cells specialize in detecting light, and red blood cells specialize in carrying oxygen. The retina too comprises cells of not one, but many specializations, organized into layers, as below.

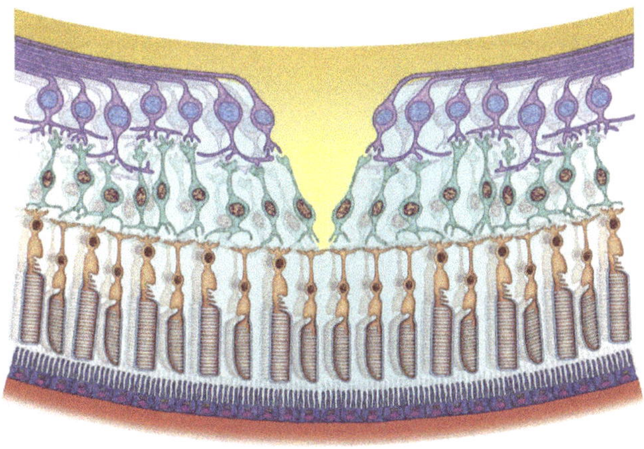

Three layers of carefully arranged cells in the retina.

The deepest layer comprises rod and cone cells which detect light and convert it to electrical signals. One would have expected these cells to be closest to the surface; strangely enough, they are placed deep inside. The next layer transforms electrical signals coming from these rods and cones and extracts useful features like contrast. Finally, the layer closest to the surface transmits this transformed signal to the brain.

Of course, this variety of cells, laid out so carefully, is obtained by repeated cell division starting from a single cell. As division progresses, the differentiation program starts taking hold. It drives different cells toward different specializations by turning on or off recipe execution on different sets of genes. Some of these cells are driven toward rods and cones, some others are driven toward cells in the middle layer, and yet others toward cells in the top layer. This happens over multiple generations of cell division. And not all these cells survive.

Much like the webbing between our fingers, nature creates sophisticated architecture by first generating more cells than it needs and then inducing programmed death in carefully chosen cells. The cells which survive, now fully specialized, enter a sort of "cellular nirvana", and escape from the cycle of repeated division. Brakes are put on the cell division program in these cells, so they no longer divide. They are then frozen as they have to hold the architecture needed for the retina to function.

This is how our retina is formed by the careful interplay of division, differentiation and death programs. Interestingly, *RB1* functions at the junction of all three programs. Just as its singly-marked form keeps the brakes on division, letting go when it is multiply-marked, its unmarked form controls parts of the differentiation program.[94]

RB1 is thus a master orchestrator, adjusting the balance between our three contestants one way or the other depending upon the need of the hour. What happened to this balance in *Ali*'s cells with two hits in *RB1*? Which way did the balance get tilted in these cells?

Ali's Cancer

Scientists have studied what happens to the various cell types in the retina when the *RB1* protein is completely removed. Many cell types continue differentiating and enter their state of cellular nirvana, escaping from the cell division cycle. The death program consumes many others. The fate of the rare cell hangs in the balance though.

In such cells, the death program is overwhelmed by the differentiation and the division programs. The differentiation program proceeds to an extent, but is in turn, overwhelmed by the division program. Only very recently have these cells been pinned down;[103] they are partly differentiated versions of cone cells, the cells responsible for sensing color.

Only in these pre-cone cells is the fine balance between division, differentiation and death upset, making division predominant. With no *RB1* protein at all, there are no brakes on division, so it proceeds at reckless speed. Cancer follows soon. Other cell types, in the retina or in the rest of the body, provide a more hostile environment for cell division in this three-way battle, so tumors only form starting from these pre-cone cells.

We can only guess, but it is likely that something similar happened in *Ali*. He inherited a first hit. Then some pre-cone cell in his retina lost exon 25 from both gene versions due to a second hit. These hits forced the *RB1* protein to concentrate outside the nucleus in this cell, and its daughter cells. All three programs—division, death and differentiation—were perturbed as a consequence. The battle among the three was then won by division. And this happened not just in one eye, but in both eyes, leading to bilateral tumors.

Had the second hit just waited for a few years allowing time for these pre-cone cells to differentiate and attain their cellular nirvana, *Ali* could have escaped retinoblastoma altogether. But the odds of the second hit, being as high as they are in the presence of an inherited hit, ensured that this was not to be.

Wrapping Up

The genomes in our various cells are not static and frozen; on the contrary, they are eminently subject to change. Indeed, *Ali*'s challenges stem from the combination of an inherited variant and some somatic variants, the latter acquired by some of his cells after conception.

Knudson's observations, made several decades ago, provided us with some insights on how these somatic variants arise—in a random manner, like a dart-player aiming his darts at cells in the retina, with an occasional chance of success. And two hits in a single retinal cell suffice for retinoblastoma to take root. Even without a hereditary predisposition, just sheer chance causes two hits in one of the millions of retinal cells, in one in 16,000 children. Luck, therefore, plays a key role in retinoblastoma.

The analysis of Knudson's numbers also threw up several mysteries. Why were the dart-players' chances of success on the first hit much higher than on the second? And why did their chances of success on the second hit appear to fade away after three years of age?

Answers to these questions stem from a comforting note—it is not easy for a normal cell (without any inherited hits whatsoever) to turn cancerous. This cell must acquire two hits for this to happen. The first is simply the introduction of a somatic variant in the *RB1* gene. The second is a more complex event though: it encompasses the introduction of another somatic variant, this time in the other version of the *RB1* gene, as also the subsequent escape from the clutches of programmed death and programmed differentiation, leading to cellular nirvana. Together, these pose a greater hurdle to our dart-players than just the task of inducing a somatic variant; hence, possibly, the lower chances of success on the second hit.

Unlike cell division in the retina, which stops before birth, differentiation continues for some time. However, after the first few years of life, most surviving retinal cells differentiate fully and attain their cellular nirvana, so chances of success for the second hit fade away with time.

In *Ali*'s case, the first hit was on account of a missing chunk comprising exon 25 of the *RB1* gene. He had inherited this hit from one of his parents, and teasing it out took a new method which we hadn't employed in any of our previous stories. The second hit was a somatic variant acquired after conception by some of *Ali*'s cells. Possibly, this second hit mimicked the first one, on account of a random crossing-over event.

Fortunately, *Ali*'s condition was diagnosed early enough to save his life, before the cancerous cells could spread from his retina to other parts of his body. Unfortunately, one of his eyes, the one in which the tumor had progressed further, had to be removed. Cancerous cells in the other eye were killed by treatment, while retaining vision. *Ali* now goes to a regular school and can do much of what other children can.

Mysteriously, neither of *Ali*'s parents had reported suffering from retinoblastoma. This was odd, for *Ali* had inherited his first hit from one of his parents. Investigations revealed that his father, without even realizing it, had what is called a *retinoma*, a tumor that had started to form and later regressed. Maybe, the odds of the random second hit, or some randomness in the balance between division, death, and differentiation, had just worked in his favor, as it does for a small percentage of individuals who have a problematic inherited hit.

If *Ali*'s parents were ever to have another child, they know what to do. They would test whether this child, while still inside the mother's womb, lacks exon 25, or not. Even if it did, there wouldn't be cause for despair. Much of the development of the retina happens at the tail end of pregnancy. Simply by delivering this baby a few weeks pre-term and starting treatment promptly,[104] any tumors in the retina can be caught early and halted in their tracks, saving both sight and life.

Story 8

Repair out of Repair

In this story, we meet eight-year-old *Toto*, who appeared to be much less energetic than others his age. Several tiny purple spots had developed on his skin, caused by bleeding in the small blood vessels just below. His six-year-old sister too had started to show similar symptoms. A blood test on *Toto* showed that his blood cell counts were much lower than normal. After eliminating several possible causes of these phenomena, *Toto*'s pediatrician performed what is called a *chromosome breakage* test to verify her only remaining hypothesis.

Remember chromosomes—those 46 books which together comprise the genomic bookshelf. The chromosome breakage test involved taking a few cells from *Toto*'s blood and inducing these to grow and divide in a dish in the laboratory. These cells were then subjected to certain drugs and observed under a microscope. Normally, each of the 46 chromosomes would appear as in the picture below.

A normal chromosome and its copy.

This picture shows just one of the 46 chromosomes. Two entire copies have been made of this chromosome as part of the genome copying process which takes place when cells divide. The two copies are connected at the center. Eventually, the copies will separate and make their way into distinct daughter cells. Forty-six such chromosome pairs would comprise the normal picture.

Some of *Toto*'s chromosome pairs indeed matched the expected picture above. Not all though—some were odd, as if multiple pairs had fused together, as shown below.

Multiple chromosome pairs fused in Toto's sample.

As the test name indicates, the drugs used in the chromosome breakage test cause breaks in chromosomes. Believe it or not, cells from normal individuals usually self-repair these breaks quite well—a boon indeed, as we shall see. In contrast, *Toto*'s cells seemed to have made a mess of it, jumbling up some of the chromosome pairs in the process. A Swiss pediatrician, Guido Fanconi, was the first to identify this condition, as early as 1927. *Toto*'s condition now carries his name: Fanconi anemia.

Anemia, or too few oxygen-carrying red blood cells, was the cause of *Toto*'s fatigue. The purple clots under his skin were caused by tiny bleeds he had developed because he had too few platelets.

Both red blood cells and platelets, as well as white blood cells, are manufactured in the bone marrow, a blood cell-making factory located inside our bones. *Toto*'s symptoms indicated that his bone marrow was starting to fail. If this continued, it could turn life-threatening. For instance, *Toto* could be at risk of bleeding to death on account of a slight bruise, or suffering from the onslaught of repeated infections, or suffering organ failure on account of decreased oxygen supply to his various organs.

Doctors could stimulate *Toto*'s bone marrow to produce more red blood cells and platelets and thus hold the fort for some time. In the medium term though, *Toto* might need blood transfusions, or even a bone marrow transplant, much like *Sita*, the girl with thalassemia we met earlier. A successful transplant would give *Toto* a reprieve. But time could present further challenges.

There was a 30–75% chance that one of *Toto*'s cells would spin out of control and turn cancerous by age 45.[105] In comparison, the average person's risk by age 45 is under 5%. *Toto*'s risk was 6 to 15 times higher. What in *Toto*'s genome caused his bone marrow to fail and his risk of cancer to shoot up dramatically? And how was that related to his chromosomes appearing jumbled up under the microscope?

Inheritance and Cancer

A brief digression on inheritance and cancer before we return to *Toto*. The risk of cancer increases with age for the average person. Interestingly, this increase in risk with age varies with gender—women carry a higher risk between the ages 30 and 60, while men more than catch up in the 60s and 70s. And much of this early increased risk for women is due to breast cancer.

Statistics indicate that 1 in 8 or 9 women will get breast cancer in their lifetime. Most of these women will not have a strong family history of breast cancer, indicating the lack of a clear hereditary cause; the disease seems to come out of nowhere in these women, just as it did for the unilateral retinoblastoma cases in the last story.

But there are exceptions.

As early as the 1940s, doctors had observed that 16% of women with breast cancer had a strong family history.[106] As more and more patients were observed and the cancer risk in immediate relatives of breast cancer patients was carefully quantified,[107] the hypothesis of familial risk grew stronger. The risk for a woman appeared higher if an immediate relative had breast cancer. And it was much higher when the immediate relative got cancer at an earlier age rather than later,[107] suggesting a hereditary component to breast cancer.

A 17-year marathon effort was conducted by noted researcher Marie-Claire King and her colleagues at the University of Berkeley from 1974 to 1990 to verify this hypothesis. In that era, comprehensively scouring the genomes of families afflicted with breast cancer was impossible. Even getting a few glimpses into their genomes was painstakingly difficult. No encouragement came from the fact that conventional wisdom of the day doubted the presence of genes in the genome predisposing individuals to cancer; particularly, cancer that sets in after a few decades of life. Nevertheless, this marathon effort eventually bore fruit by providing incontrovertible evidence that variants in a specific gene on chromosome 17 increased the risk of breast cancer substantially.[108]

The gene itself was identified only a few years later, in 1994. It is, arguably, the gene with the greatest presence in the public arena among all the 20,000 or so genes in our genome. It is also connected, albeit very indirectly, to *Toto*'s travails.

The *BRCA* Genes

This gene was appropriately christened the *BRCA1* gene—pronounced *braca one*, short for breast cancer one, early onset. Shortly after its discovery, another cancer susceptibility gene was identified by Michael Stratton and his colleagues at Cancer Research, UK, in 1995, and named along expected lines: *BRCA2*.[109] Roughly 60% percent of women who inherit a problematic *BRCA1* variant will develop breast cancer by age 70. Likewise, for 45% of women who inherit a harmful *BRCA2* variant. Compare this with the average woman, whose lifetime risk is only 12%.

So significant is this increased risk that women carrying problematic *BRCA1* or *BRCA2* variants are offered the rather drastic choice of risk-reducing surgery: preemptive removal of their breast tissue. Of the many women who have exercised this option, actress Angelina Jolie's case is the most-publicized.

In an open letter in the New York Times in 2013[110] that created much awareness about the *BRCA* genes, Jolie stated that her mother had fought cancer for almost a decade before she succumbed to it at age 56. Jolie had a problematic *BRCA1* variant and decided to take the drastic step of risk-reducing surgery to preempt that tragic fate. As a result, her risk of breast cancer reduced from the 85% number down to about 5%, as estimated by her doctors. She continued to carry a 55% risk of developing ovarian cancer, for problematic *BRCA1* variants predispose one to both breast cancer and ovarian cancer (and a few other related cancers). Women with *BRCA1* and *BRCA2* variants are indeed offered the option of surgically removing their ovaries as well, possibly after expediting plans to have children. A little later, Jolie announced that she had indeed undergone elective surgery to remove her ovaries and fallopian tubes as well.[111]

Given these choices, testing of women for *BRCA1/2* gene mutations became a topic of much business interest from the late 1990s onwards. This, in turn, led to one of the most intensely controversial patenting questions of all time—can genes be patented?

Patents for the *BRCA1* and *BRCA2* genes were granted to a company called Myriad Genetics and its academic partners in 1997 and 1998, respectively. This provided Myriad Genetics with the exclusive right to test women for problematic *BRCA1/2* mutations, blocking all other companies from doing so. Arguments and counter-arguments were made for the validity of this patent over the years. Genes were, after all, natural entities on the one hand, which cannot be patented. On the other hand, Myriad was running its test not on the gene itself, but on a copy of the gene, a modified entity which it argued, was a man-made invention. What extent of modification qualifies for patentability then became the moot-point. Meanwhile, sequencing costs for these two genes dropped to a few hundred dollars, adding to much public opinion against these patents. Having received some adverse court rulings, Myriad is now settling its disputes with its competitors, paving the way for broader access and lower costs.

But what do the *BRCA1/2* genes have to do with *Toto*'s condition? These genes have alternate names: *FANCS* and *FANCD1*, respectively. In both cases, the prefix *FANC* stands for Fanconi anemia. The same Fanconi anemia that *Toto* was afflicted with.

Genomic Assaults

So, coming back to *Toto*, *FANCS* and *FANCD1* are not the only genes with the *FANC* prefix, there are several others. Variants in these genes are known to cause Fanconi anemia. And these genes have a common role: to protect the genome from various damaging onslaughts to which it is constantly subjected.

Yes, the genome comes under constant attack by various stimuli in its environment. Some of these onslaughts occur only occasionally: for instance, radiation from X-rays. Some, however, are far more regular and routine—every minute, in fact.

Among the most routine tasks our cells do is the generation of energy and packaging of this energy in batteries called ATP for use whenever needed. Remember, we encountered this battery in *Dia*'s story on retinal blindness. The charging of this battery takes place over a series of reactions in little organelles called *mitochondria*. Each reaction grabs electrons from one molecule and gives it to the next, releasing energy in the process. This cascade terminates when the final electron is absorbed by oxygen to yield water. The energy released in this entire process is used to turn a molecular motor, which then coerces an extra phosphate unit on a molecule called ADP to generate ATP, thus charging the battery.

So far, so good. What is problematic is the small fraction of electrons that escape from this cascade prematurely and combine with oxygen to yield some unexpected molecules; for example, negatively charged oxygen, or *superoxide* as it is called. This version of oxygen is highly reactive and toxic to our cells. Fortunately, our genomes have gene recipes that help convert most of the superoxide to less harmful forms like hydrogen peroxide. Hydrogen peroxide easily diffuses out from the mitochondria to all parts of the cell, where much of it is converted to water and removed. Some, however, is converted to other highly reactive and toxic forms. These toxic forms of oxygen, collectively called *ROS*, short for *reactive oxygen species*, then react with our genomic characters.

"Can characters react?" you might ask. Well, genomic characters are not just plain characters, as the previous stories might have had you believe. The picture below, even if arcane to the non-chemist, should convince you that these are actual molecules.

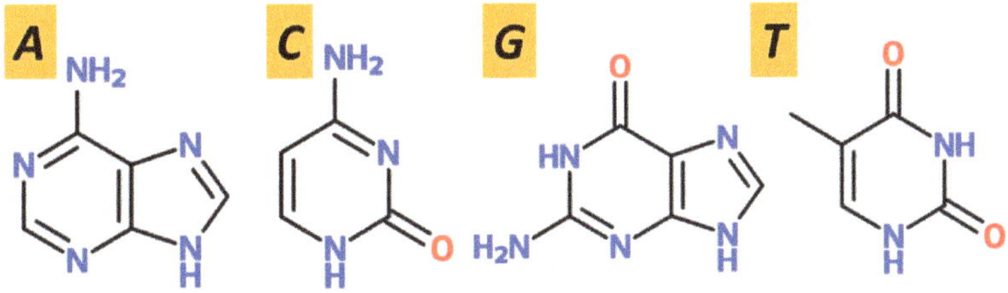

Molecules corresponding to the four genomic characters. C and T look somewhat similar to each other, as do A and G.

These molecular characters can indeed react with ROS and get modified in the process. For instance, reaction with ROS comprising an oxygen and a hydrogen atom can sometimes convert a G to a form called OH8-G as shown below.

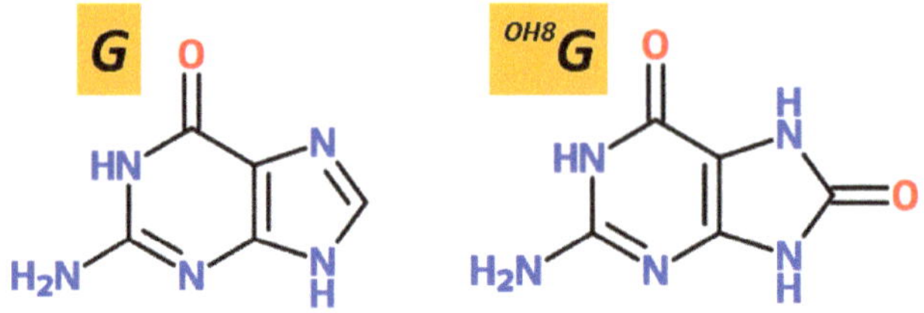

The bona fide G, and OH8-G with its extra oxygen and hydrogen.

The problem with this unusual G is, of course, that it behaves differently. To understand why, we need a brief digression. So far, we have portrayed the genome rather simplistically as a sequence of characters A, C, G, and T. In reality, each instance of the genome comprises, not one, but two character sequences. These *strands*, as they are called, are *complementary* to each other; if there is an A on one strand, there is a T on the other, and if there is a G on one strand, there is a C on the other, as you can see in this picture.

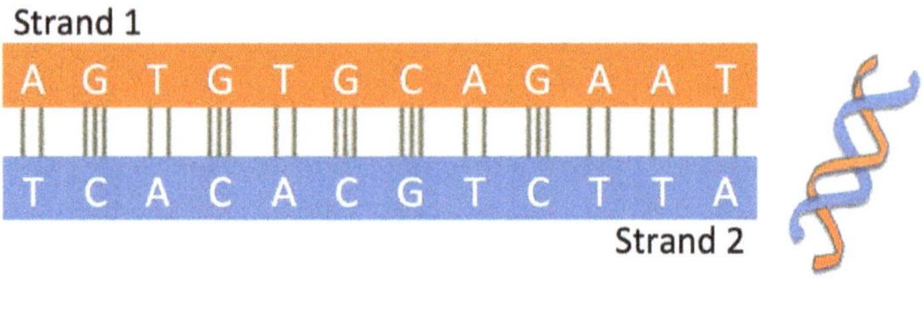

The two complementary strands of the genome. Note, these are not two versions of the genome; rather they are two strands within *the same version. In reality, these two strands are intertwined to create the iconic double helix form shown on the right.*

We usually focus on just one of these two strands; the other is perfectly complementary and often not worth dwelling upon. On occasion though, this complementarity is disturbed. For instance, while G usually pairs with a C on the other strand, the unusual OH8-G can occasionally pair with an A, thus breaking the standard rules. This has serious consequences, as the following scenario illustrates.

Imagine that we start with a C–G pair: a proper C on one strand and the expected G on the complementary strand. Then, suppose reactions with ROS convert this G to OH8-G, as shown below.

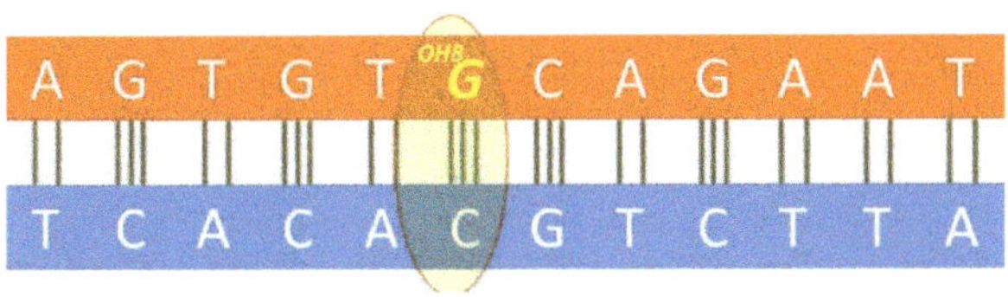

ROS convert a G to OH8-G.

Subsequently, imagine the cell divides into two. A copy of the entire genome is made in this process. Toward this, the two strands are unzipped first by breaking the bonds that hold them together, much like unzipping a zipper. Once done, brand new complementary strands are created, as in the picture below. This yields two copies of the genome, one for each daughter cell.

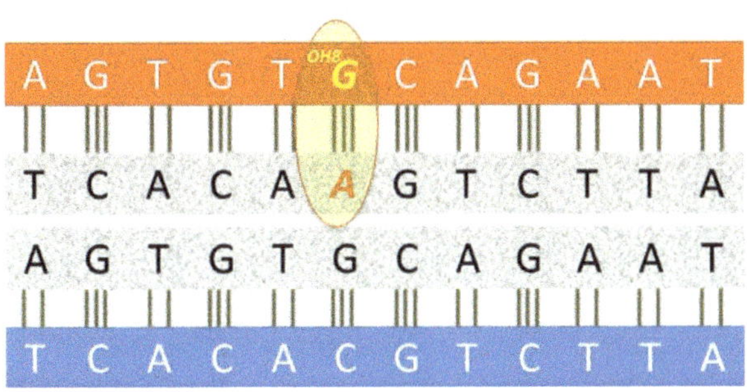

How the genome is copied. The two original strands are unzipped and the two brand new strands are shown shaded.

If you held an alert eye, you would have noticed something strange in the first copy in the picture above. Rather oddly, the aberrant OH8-G has been paired with an A instead of the customary C.

What happens next when the daughter cell inheriting this copy divides? Again, the two strands in this copy are unzipped and new complementary strands are synthesized, yielding two copies of the genome, one for each granddaughter cell. One of these granddaughter cells will now have an A–T pair where there should have been a C–G pair. And all her descendants will carry this A–T pair, blissfully unaware that life at this position in the genome began with the C–G pair, which mutated somewhere along the way on account of ROS.

ROS can cause other character changes as well. If you carefully look at the molecular characters shown a little earlier, you will notice that the molecule for character C looks similar to that for character T. Indeed, ROS can convert the occasional C to a form close to T. Subsequently, when the genome is copied, this character actually does become a T. Going further, ROS can even cause characters to be removed from the genome altogether.

Thus, as cells divide, daughter cells acquire genomes that may differ from the genomes in their respective parent cells. These *somatic* mutations, triggered by ROS, accumulate with age. When enough of these accumulate in a single cell, things could start going wrong, as we saw in the story on *Ali*'s retinoblastoma, and as we will see in more detail in the next story.

Note the rather pedestrian nature of these ROS—they are the by-products of routine, everyday life. Our cells do have protective enzymes which neutralize these ROS. Nevertheless, sufficient numbers escape to create 10,000–100,000 genomic hits per day per cell![112] Adding fuel to this fire are additional insults: cigarette smoke, asbestos fibers, diesel exhaust, fine particles in the air,[113] ultraviolet light,[114] and various other pollutants. All of these add to the risk substantially.

Toto's genome was no doubt exposed to all these onslaughts. But so was everyone else's genome. What then made *Toto* different?

Repairing the Damage

When storage disks on a computer fail, say on account of a scratch or a fault, the computer itself fails to function properly. So would parts of *Toto*'s system, if the genome in certain cells in his body were to develop scratches and faults. Fortunately, unlike computer disks, our cells can repair these genomic lesions themselves! Well, at least to good measure.

For instance, suppose a G becomes OH8-G on reaction with ROS. Our cells will promptly detect this change. Several proteins will then be set upon the task of correcting this error, in two steps. First, the erroneous characters, along with a few surrounding characters, are snipped off by these proteins. This leaves a gap in the genome as shown here.

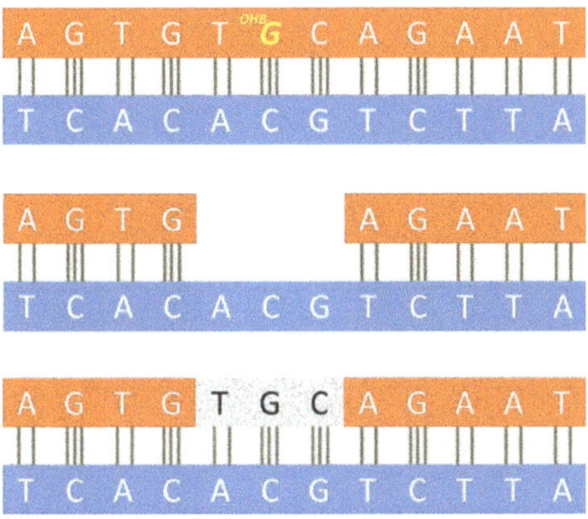

Repairing a defective character using the other strand. The missing characters TGC are filled in based on the complementary characters ACG on the other strand.

Other proteins then come in to fill this gap. However, they face a challenge: which characters should they fill into this gap? This is where the double-stranded nature of the genome helps. If characters are missing from only one of the two strands, the other strand helps determine which characters to fill in. For instance, if that strand has a C, then the missing character must be a G, and so on. The gap is then filled in using this intact strand, as shown above.

And that raises the next question—what if the opposing characters on both strands were to be damaged? No longer can one strand be used to fill in the other. How would our cells know what characters to fill in then? Were *Toto*'s cells equally adept on this count?

Crosslinks and Genomic Surgeries

Here is one such scenario of particular relevance to *Toto*. Usually, a G on one strand is bonded with a C on the other. This bond is strong enough to keep the two strands together but is easily broken when the strands need to be unzipped for genome copying. Occasionally though, a much stronger link can form between a character on one strand, say a G, and not quite the complementary C but some other nearby character on the other strand, say a T. These two characters get connected together, as shown below, by what is called a *crosslink*.

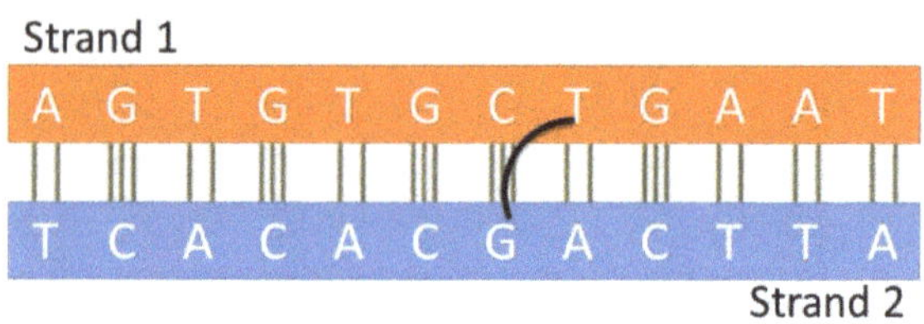

Crosslinks in the genome.

This crosslink is so strong that it prevents the two strands from unzipping completely when the genome is being copied. The genome-copying process then stalls, placing the entire cell division process under suspension. To make progress, this crosslink must be removed. Our cells are up to this challenge, but using a very elaborate procedure,[115] described next.

Let's start with a situation where the two strands have been unzipped and copies created on both sides of the crosslink, as shown in this picture. This process is now stalled and cannot proceed to completion, thanks to the crosslink.

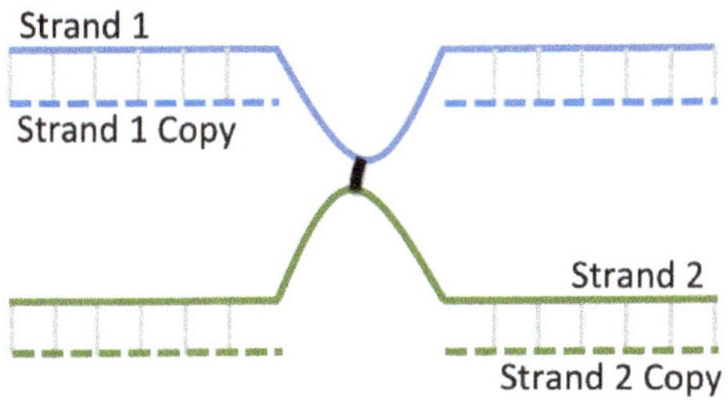

The crosslink stalls genome copying.

As before, special proteins detect that a crosslink has stalled the copying process and launch an elaborate operation. First, a short stretch of characters around the crosslink is snipped off from one of the two strands. This stretch completely detaches from this strand, but remains attached to the other strand because of the crosslink, as in the picture below.

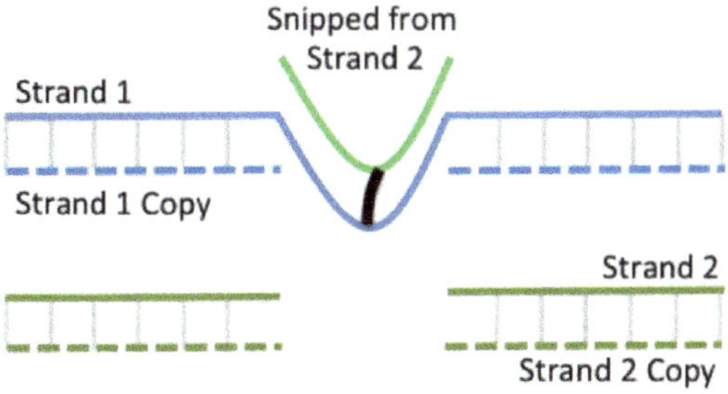

A stretch containing the crosslink is unhooked from Strand 2.

Strand 1's copy is now ready to be completed. Since Strand

1 itself is still intact, this is easily done by adding complementary characters, as shown below. The snipped-off portion is eventually dropped as well, using a similar process of snipping-off and filling-in.

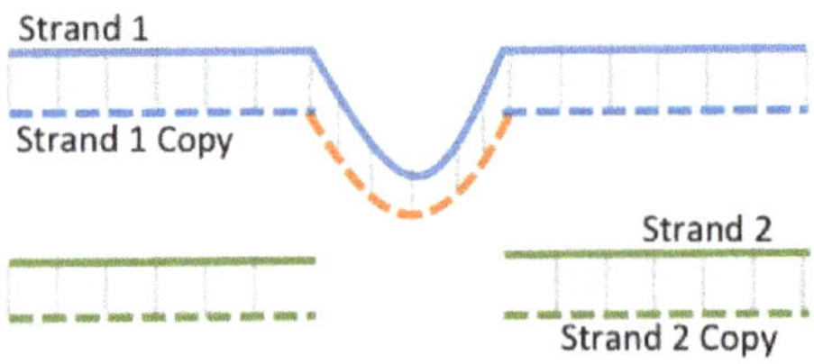

The copy of Strand 1 is now complete.

Completing Strand 2 now poses a challenge: what does one fill into the missing region? Neither Strand 2, nor its copy, have any memory of these characters. The only place where this information is available is in Strand 1 and its copy, which is now complete. But Strand 1 and its copy together constitute a separate, very large molecule. And the information required to fill in the missing region in Strand 2 is buried somewhere deep inside that molecule and needs to be fished out. How do our cells hunt for this information?

Indeed, our cells have evolved clever ways to perform this search. Rather strangely, they begin by removing some more portions from Strand 2 and its copy, as shown in this picture.

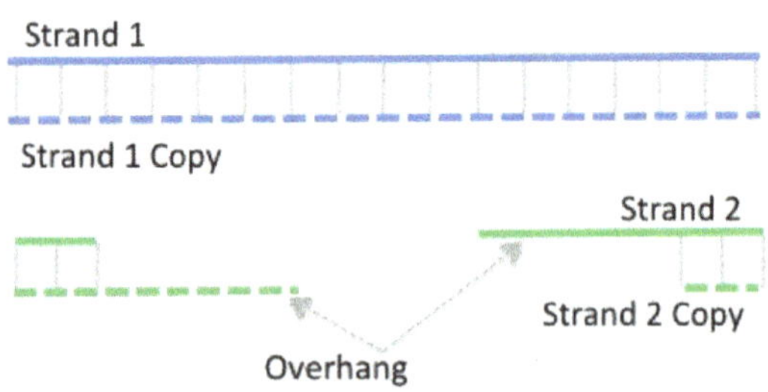

Some more portions of Strand 2 and its copy are removed.

Momentarily, the removal of good stretches from Strand 2 and its copy adds to already existing damage. However that is part of a larger game plan of repair. The overhanging portions of Strand 2, now stripped of their complementary mates on the opposite strand, seek to find these mates again. With the assistance of some helper proteins, they probe Strand 1 and its copy for stretches of characters that are complementary to themselves, as in the picture below. For instance, if Strand 2's overhang has the character sequence ATGCA, it seeks the sequence TACGT in Strand 1.

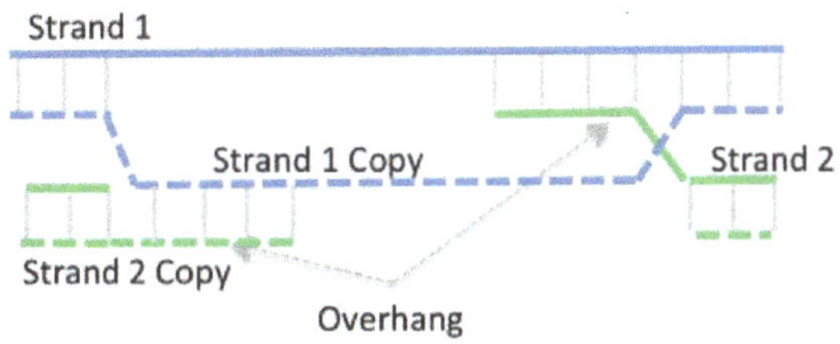

The overhangs in Strand 2 and its copy perform a search to locate complementary stretches in Strand 1 and its copy, respectively.

Miraculously, this so-called *homology search* is accomplished quite accurately by our cells. Once such a matching complementary stretch is found, the overhangs cause unzipping of Strand 1 and its copy, as shown in the picture above, allowing the overhangs to bond to their complementary stretches. The stage is now set for filling in the missing piece, as shown in the next picture.

The gap in Strand 2 is filled in using the information provided by Strand 1. To this end, Strand 2's overhang is simply extended with characters that are complementary to Strand 1. Likewise, the overhang in Strand 2's copy is extended with characters that are complementary to Strand 1's copy. Once both extensions are done, we have two full copies of the genome.

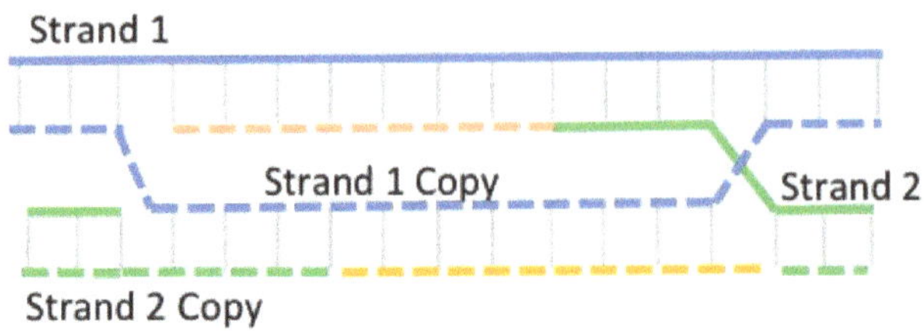

The overhangs are now extended using the complementary strands.

One final task still remains: that of disentangling the two copies, for they are rather strangely entangled at the moment. There are different ways in which this surgery is performed, but one common way ends up cutting the various strands so it yields the final result shown in this picture.

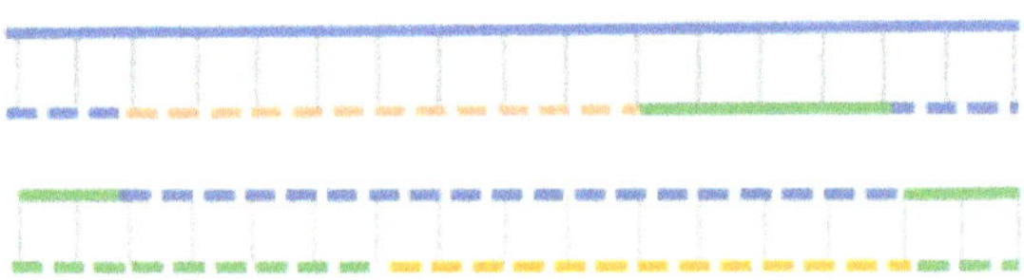

Both strands are fully copied now.

So many surgical steps, just to fix a little crosslink! It would have been far simpler had nature evolved a direct way to break and dissolve these crosslinks. But all we have is this more convoluted multi-step surgical route.

This surgery has to be carefully shepherded by proteins obtained from a number of genes. Over the years, several individuals have been observed with problematic variants in these genes. In these individuals, the above surgery doesn't quite go according to plan.

When a drop of blood is taken from such individuals and subjected to drugs that induce high levels of crosslinking in the genome, the outcome can be surprising: different chromosomes are mistakenly stitched together, sometimes in pairs, sometimes even triplets—the disorderly aftermath of several botched surgeries. Indeed, *Toto*'s chromosomes too presented such a chaotic picture. Clearly, one or more of the genes responsible for careful genomic surgery was malfunctioning in *Toto*, leading to Fanconi anemia. Several such genes are known. Like the *BRCA1* and *BRCA2* genes, many of these carry aliases which begin with the prefix *FANC*, short for Fanconi anemia.

The *FANC* Genes

The list of *FANC* genes is long and neatly labeled in alphabetical order: *FANCA*, *FANCB*, *FANCC* and so on; 17 and counting at the moment. Problematic variants in each of these genes have been found in patients with Fanconi anemia. For brevity, we'll drop the common *FANC* prefix and refer to these genes just as *A*, *B*, *C* and so on. Most letters are used only once, but the occasional letter may be taken by two genes, for instance, *D1* and *D2*.

Together, these genes, along with several others, work in tandem to carry out their mission: detecting certain types of damage to the genome and orchestrating the careful surgery required to fix this damage.[116] This mission of repair is initiated when the *FANCM* protein (just *M* for short) senses that genome copying has been stalled by a crosslink.[117] It promptly signals for help, recruiting its brethren *A*, *B*, *C*, *E*, *F*, *G* and *L* into the mission. These proteins band together to form a combined structure, called the *core complex*.

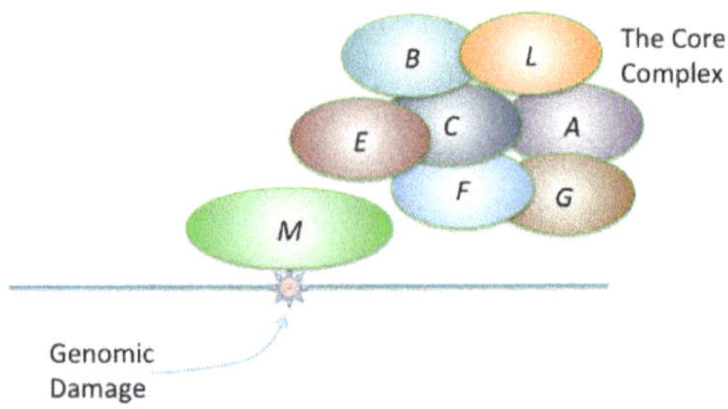

M senses the damage and recruits the core complex.

This core complex then primes its emissaries, the *D2* and *I* proteins, by marking them with a special mark. This mark is actually a small protein called *ubiquitin*. This mark is delivered specifically by the *L* protein in the core complex, as shown below.

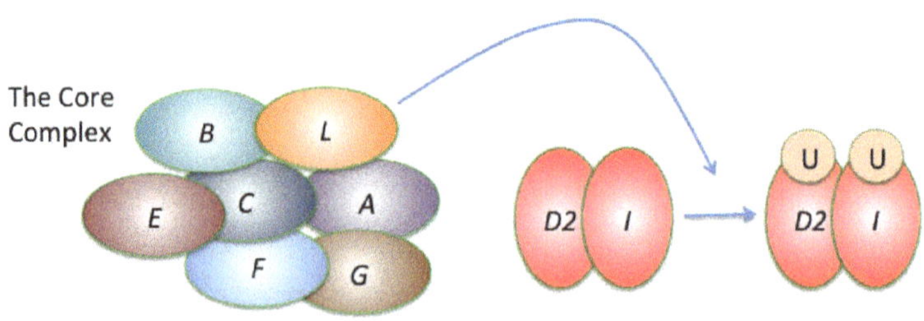

The D2 and I proteins, primed by a ubiquitin mark put by the L protein.

Once marked so, the *D2* and *I* proteins locate near sites which require attention in the genome. Here, they set about their task of recruiting various proteins required for the multi-step genomic surgery shown in the earlier pictures. These include proteins that snip the genome, proteins that fill in missing pieces, and proteins that help perform the homology search and subsequent steps required for filling in the missing piece.

Some of these proteins are shown in the picture below. Detailed roles of these proteins still need to be elucidated, but some rough outlines are known: for instance, *O* helps recruit and manage proteins which attach to the overhangs and assist in the homology search. Note the *BRCA1* and *BRCA2* proteins also appear in this list[118] (as *S* and *D1*, respectively).

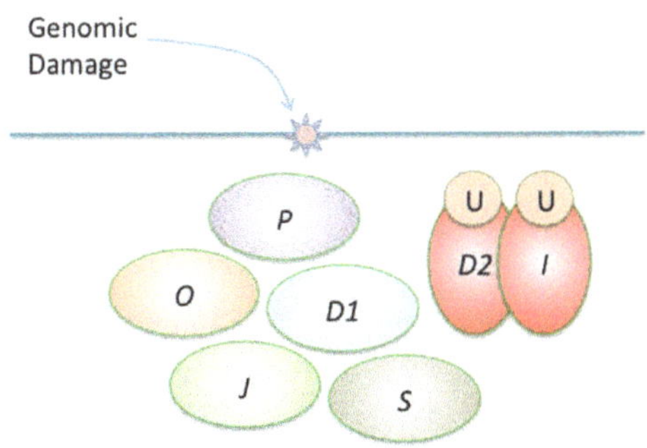

The other FANC proteins, which coordinate the repair process.

Once the surgery is completed and the damage repaired, the ubiquitin markers are removed from the *D2* and *I* proteins, indicating that they are no longer busy. They then wait for the core complex to set them busy again, in response to the next assault on the genome, and so on.

This is how the *FANC* proteins keep up their vigil, ensuring the genome is protected. In *Toto* though, one or more of these ever-vigil proteins had relaxed its guard, thus leaving the genome unattended and exposed to an accumulation of uncorrected damage. Which gene's recipe did that protein come from?

The Hunt for the Variant

The cause of *Toto*'s challenges was likely a problematic variant in one of the 17 known *FANC* genes. More *FANC* genes causing conditions like those of *Toto* are still being discovered; nevertheless, as a first measure, we focus on just these 17 genes.

As in the last story on *Ali*'s retinoblastoma, we sequence all exons of roughly 100 genes, all well known to carry problematic variants that increase the risk of cancer in an individual. The 17 *FANC* genes are included in this list. Of course, when we sequence an exon, we also sequence 50–100 characters into the neighboring introns.

Note that both of *Toto*'s parents are healthy; they do not have any of the problems that *Toto* or his sister have. *Toto*'s Fanconi anemia thus appears to be a *recessive* disease—both versions of a relevant gene must have been rendered dysfunctional by problematic variants in *Toto*. Each of his parents probably has just a single dysfunctional gene version and is protected from disease by the other good gene version. So, we look specifically for problematic variants present in both gene versions in *Toto*—either the same variant present in both gene versions, or two distinct variants, one in each gene version.

One *FANC* gene presents an exception to this rule, though. The *B* gene is actually present on the X chromosome, of which *Toto*, being male, has only one version. As a special case for this gene, we look for variants present in its sole copy in *Toto*.

Our search, as has often been the case in these stories, hits a roadblock at the very outset. At first sight, there appear to be no variants of the types we are seeking in *Toto*'s genome. The only variant of note, if any, is an innocuous-looking G-to-A variant in the *FANCL* gene. It appears to be clearly innocuous, for it causes no change at all in the recipe triplets.

A quick summary again, in case you need to be reminded of these triplets. Remember how recipes are obtained by grouping characters into triplets; each triplet is then converted to an amino acid. The next picture below should refresh your memory. In this picture, our G-to-A variant causes its triplet to change from AAG to AAA. Both triplets are *synonymous*—they describe the very same amino acid, namely, *Lysine*, whose short form is *K*. Effectively, the recipe of the *FANCL* gene does not change at all. With no change to the recipe, it is unlikely that this variant can cause *Toto*'s rather extreme condition.

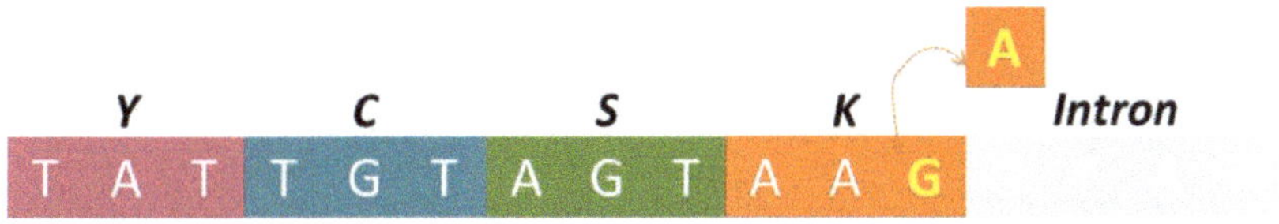

A synonymous variant in exon 13 of FANCL. Above the triplets are the corresponding amino acid short forms.

Other than this variant, there appears to be little of note. None of the usual characters in the *FANC* genes seems to have been done away with in *Toto*. No extra characters that shouldn't be there are present. No large chunks appear to have been disposed of or put in. There is nothing that might truncate any of the *FANC* gene recipes prematurely. There are no variants that cause a recipe change from one amino acid to another. There aren't any variants in the overhangs into the introns either that might mislead recipe execution as it jumps from exon to exon. In summary, there is absolutely nothing of significance that might help us explain the cause of *Toto*'s singular condition.

Accordingly, we send a note to *Toto*'s doctor stating our inability to identify the offending character in *Toto*'s genome. Pop comes back the challenge. *How can that be? The chromosome breakage test was crystal clear, putting the diagnosis of Fanconi anemia well beyond doubt. There must be a variant responsible for this in Toto's genome. Please look harder.*

The Exonic Cliff

Hectic team meetings follow in the face of this renewed challenge. The only variant we have to work with is the G-to-A change in the *FANCL* gene. This variant causes no change to the gene recipe though, at least at first sight, so there is little evidence to its culpability. Then, someone notices a curious fact.

As the picture above shows, this variant appears at the very end of exon 13, after which the genome slips off into the intron that follows. Perched precariously at the edge of this cliff, could this variant throw the jump to the next exon off-balance during recipe execution?

We review these jumps from *Sita*'s thalassemia story briefly in order to attempt an answer to this question. Remember, recipes are described only in the exons. When executing the recipe written in these exons, intervening introns are skipped over. One cannot but marvel at this process which launches from the end of an exon and lands precisely at the beginning of the next exon, leaping across thousands, if not tens of thousands of characters in the intronic ocean. How does the recipe execution process know where an intron begins and where it ends? This picture should refresh your memory.

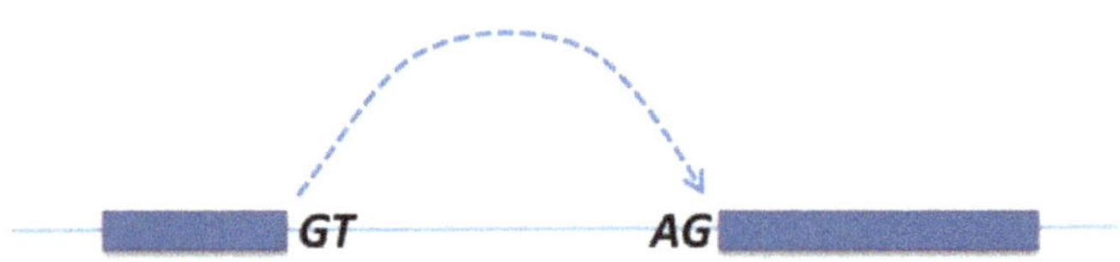

Recipe execution jumps from the GT marking the end of an exon to the AG marking the beginning of the next.

These characters, GT and AG, inform the recipe execution process where to take off and where to land, respectively. Of course, these are not the only guides, for there are GTs and AGs galore even in the middle of exons and introns. As we saw in *Sita*'s thalassemia story, the landing of the trans-intronic jump is guided by additional characters: the branch point and the C/T stretch, both of which appear close to the end of the intron. But how about the take-off? What additional characters guide the take-off? Does the last character of an exon have a say in the matter?

To answer this question, hundreds of thousands of exons and introns from all 20,000 or so genes are taken and their character sequences juxtaposed to identify which character patterns occur commonly at the exon–intron boundary. The picture below depicts this pattern.

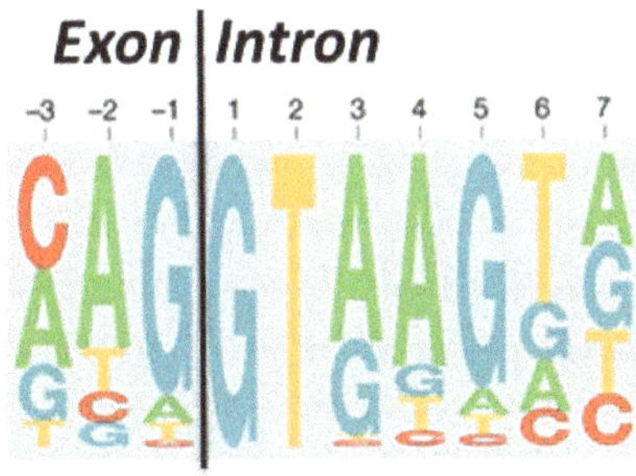

Character prevalence around the exon–intron boundary.[119]

As this picture shows, most introns have G and T, respectively, as their first two characters; hence the full-sized G and T at positions 1 and 2. This overwhelming prevalence underscores their importance in guiding the take-off of the trans-intronic jump. The smaller sizes of the other characters indicate that they are less preserved across hundreds of thousands of exons and introns, and hence their contribution to guidance is less clear. Nevertheless, if you look at position -1 in the picture, you may see a pattern of note.

Position -1, the last character of an exon, is very often a G. Not always, but very often, with other characters occurring in relative minority. This is suggestive that the end of an exon needs a G to guide the take-off of the trans-intronic jump.[120] The G-to-A variant in the *FANCL* gene in *Toto*'s genome is precisely at this position—the last character in exon 13. Could this alteration from G to A in *Toto*'s genome have tricked the jump into not taking-off?

Exon Extended, Contracted, or Skipped?

Let's assume for a moment that this variant, which lies at the end of an exon in *Toto*'s genome, indeed misleads the splicing jump. What happens next? Does the recipe execution process simply continue further into the intron rather than jumping over to the next exon?

That is indeed a possibility. But there is another, less obvious, possibility as well:[120] the jump to the next exon still happens, but from a different launch point instead of its usual springboard at the end of this exon. A closer look at these splicing jumps suggests why this is a possibility.

The *FANCL* gene has as many as 13 introns, which must be removed before the recipe is executed. The character pattern in the picture above serves as a guide for this purpose. Several places in the genome might match this pattern, some better than the others. For instance, the sequence CAGGTAAGTA matches the pattern in the picture quite strongly—each character is the most likely character at that position. But CAGGT*GG*GTA also matches the pattern reasonably, albeit not as strongly, with the most likely characters AA at positions 3 and 4 in the picture replaced by GG. Several such character sequences at different genomic positions might stake their claim to initiating the beginning of an intron, some more strongly than the others. Among all these competing choices, the cell chooses only a few to launch its trans-intronic jumps. How does it pick its choices?

The rules which determine the winners of this competition are still a subject of research. One such rule is, as you might expect, that positions which show greater closeness to the pattern in the picture above have a better shot at winning.[121] There are other rules as well,[122] but none that is fully predictive of the winner. Nevertheless, a variant, such as the G-to-A variant in *Toto*, might suddenly alter the results of this competition.[120] One of the erstwhile winners might now present a poorer match to the desired pattern and an erstwhile loser might now emerge the winner. This new winner might well provide an alternative launch point. Sometimes, this launch point moves a little further down into the intron, extending the exon a bit and adding new characters to the gene's recipe. At other times, it appears earlier in the exon, effectively removing characters from the gene's recipe. On occasion, something more dramatic happens.

In such instances, the launch point moves all the way to the end of the previous exon. In other words, two introns are skipped in one jump, along with the exon in between, a super long jump at that. It is hard to predict when this happens, but we know the following. The various introns are not all removed at once, but in some order, which is not necessarily first to last. Variants near the beginning of introns that are removed earlier in the process tend to result in this super long jump, in contrast to those at the beginning of introns removed later in the process.[123]

The G-to-A variant in *Toto* could have any of the consequences listed above—some extra characters might enter the recipe, some characters might be deleted from it, and even a whole exon might be deleted altogether. How do we know which, if any, is the case?

One way to find the answer is to monitor the recipe execution process as it progresses. It first reads all the exons and introns and copies them into a new molecule. From this new molecule, all the introns are removed. If we could read the sequence of characters at this stage, we would know exactly which exons remain in this molecule and which introns are removed. That would give us our answer. Fortunately for us, someone had already performed this experiment, that too very recently.[124] The answer from this experiment is in the picture below.

Exon 13 completely skipped due to the G-to-A variant at its end.

The entire exon 13 is missing from this molecule; the recipe execution process jumps straight from exon 12 to exon 14, removing both exon 13 and the two intermediate introns because of the G-to-A variant at the end of exon 13. What appears to be an innocuous variant at first glance actually has a major, insidious effect.

Does it explain *Toto*'s condition though?

Botched Genomic Surgery

Like a single bad apple bringing down the whole pack, a single character alteration knocks all 72 characters of exon 13 out from the *FANCL* gene recipe. When grouped into triplets, these 72 characters comprise 24 triplets, numbered 341–364. These triplets are missing from both versions of the *FANCL* gene in *Toto*. How important are these missing triplets?

Remember *FANCL*'s role in fixing crosslinks in the genome? It swings into action when they are detected. It then commands its emissaries, *FANCD2* and *FANCI*, to go recruit other proteins that can perform the various genomic surgeries required to fix this crosslink. This command is issued by the addition of a special ubiquitin mark to *FANCD2* and *FANCI*. By deliberately altering specific amino acids in *FANCL* and studying the consequences of this alteration, the hand that issues these marks has been pinned down to lie in triplets 307–359. Specifically, when triplets 341 or 359 are altered, *FANCL* loses its marking ability.[125,126] And both these critical triplets are missing from *Toto*'s recipe.

With both amino acids 341 and 359 lost in both gene versions, *FANCL* probably loses all ability to mark *FANCD2* and *FANCI*. The command to go and fetch proteins that can perform corrective genomic surgery is then no longer issued in its usual manner. And what is the consequence?

Remember the missing piece in Strand 2 that was introduced when fixing the crosslink? To refresh your memory, it appears again in the picture below.

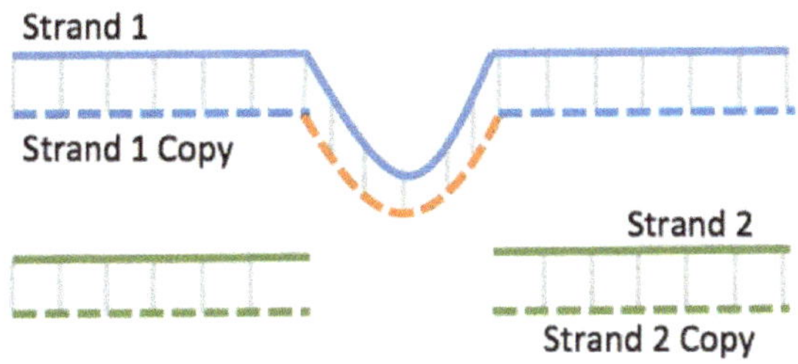

The missing stretch in Strand 2; how is it filled in now?

Filling in this missing piece correctly is a challenge for the cell. The complicated surgical maneuvers we went through earlier were all aimed at this task. These maneuvers are orchestrated by a properly functional *FANCL*. In *Toto*, *FANCL* can no longer perform this orchestration. Consequently, the proper sequence of surgeries is not invoked. Rather, the missing piece is filled in using a short cut, or a hack, as engineers would call it. Instead of filling in the missing piece by performing a homology search on Strand 1, this hack simply joins the two disconnected fragments of Strand 2, dropping the missing piece altogether in the process.

No doubt, this is a dangerous fix, particularly when the missing piece carries some critical parts of a gene recipe. But the problem doesn't stop here; sometimes, things can get even more chaotic. The hack can go further and make a greater mess by joining the two fragments of Strand 2 not to each other but to other fragments resulting from other crosslinks elsewhere in the genome[127]—sometimes, even fragments on other chromosomes. The net result of this hack: fused chromosomes of the type that *Toto*'s doctor saw when she performed the chromosome breakage test on his chromosomes.

Of course, the chromosome breakage test administered by *Toto*'s doctor used specific drugs to induce crosslinks in large numbers in a blood sample. The cells in *Toto*'s body themselves would not experience crosslinks at nearly such high levels in their natural course. However, several common stimuli do induce crosslinks in the genome, albeit at lower rates; these include metabolites of alcohol, cigarettes and high-fat diets, as well as nitrites used to preserve food.[115] Our cells usually repair most of these crosslinks without batting an eyelid. Unfortunately, *Toto*'s cells were highly compromised in this ability.

This inability to repair crosslinks probably caused faster accumulation of genomic errors in *Toto*'s cells as compared to the average person. Usually, cells detect such high levels of genomic damage and slow down their cycle of division to allow time for repair. However, when this damage is irreparable, as might happen when fragments are joined randomly as in the picture above, nature can only induce cells to eliminate themselves by programmed death—a safer option than continuing to multiply with defective genomes,[128] as you might recall from the last story. The result: cells in *Toto* had a much higher rate of death than usual.

This phenomenon is most apparent in the bone marrow, the factory located in the interior of our bones. Here, so-called *stem cells* divide continually to generate new blood cells, replacing old blood cells which are turned over once in several days. *Toto*'s bone marrow stem cells were instead undergoing programmed death in large numbers. Inevitably, this was leading to bone marrow failure, the inability to generate healthy blood cells in sufficient numbers. All due to the presence of what seemed like an innocuous A instead of the usual G at the end of the exon 13 in both his *FANCL* gene versions.

Wrapping Up

An individual begins life as a single cell, which then divides into tens of trillions of cells. Some of these cells continue to divide throughout our lives, replacing other cells that die. Over these trillions of cell divisions, the genome doesn't stay unchanged. Various routine exposures of life cause damage to its molecular structure on an ongoing basis. Then, as cells divide, damaged molecular structure forces the character sequence in a daughter cell to be slightly different from that in the parent cell. Once the genome in a cell changes, all daughter cells arising from further division of this cell inherit these changes. Over time, our body therefore becomes a genomic mosaic, with different cells carrying slightly different versions of the genome.

Fortunately, our cells have several protective mechanisms to slow down the rate at which this genomic variety builds up. Repairing damage to the genome is one such mechanism. Regardless, as the juggernaut of age rolls on, the mosaic gets more and more varied. Occasionally, enough genomic change accumulates in some cell that it cripples the recipes that enforce these protective mechanisms. This allows that cell to then divide and accumulate genomic change even faster, a condition called *genomic instability*, which in turn leads to cancer.

Of course, if the ability to repair damage to the genome is compromised, then this picture is accelerated and genomic instability sets in much earlier. Indeed, women with problematic variants in just one version of the *BRCA1* or *BRCA2* genes (also called *FANCS* and *FANCD1*) have a 70–80% and a 50–60% chance, respectively, of getting breast cancer in their lifetime.[129] However, such cancer usually sets in after a few decades of life. Presumably, the sole working gene version holds the fort for a while, until some erroneous genomic surgery cripples it—a phenomenon that happens often, but not always.[129]

Unfortunately, *Toto* had both versions of a key repair gene crippled by a subtle variant whose impact on the gene's recipe was most insidious. Each of his parents carried the variant in one version, so they were completely unaffected and would never have realized what was in store for their children. Genomic instability in *Toto*'s bone marrow cells had set in early as a consequence. The protective measure of programmed death then resulted in an acute shortage of bone marrow cells.

Toto would need a bone marrow transplant to cure this condition. If he were to successfully negotiate this hurdle, then the next challenge will be an increased likelihood of cancer as he grows older. All on account of a single character alteration in his genome, which we dismissed as triflingly innocuous at first glance.

Hopefully, time is on *Toto*'s side, and tremendous advancements in detecting and treating cancer will be made in the interim, so the world is ready to manage *Toto*'s cancer if and when it sets in. The next story will give us a glimpse of where the world is headed on this front, and how the very crosslinks that bother *Toto* are also important weapons in our fight against cancer.

Story 9

Moves & Countermoves

In this final story, our focus moves to our battle with cancer. The subject of this story is *Siva*, who was prompted to see a doctor on account of persistent abdominal pain when he was 37. An ultrasound showed excessive fluid accumulation in his abdomen. Why this fluid had accumulated was explained only when a CT scan was performed. It showed several nodules growing abnormally around the lungs, as also an abnormal growth in the abdomen.

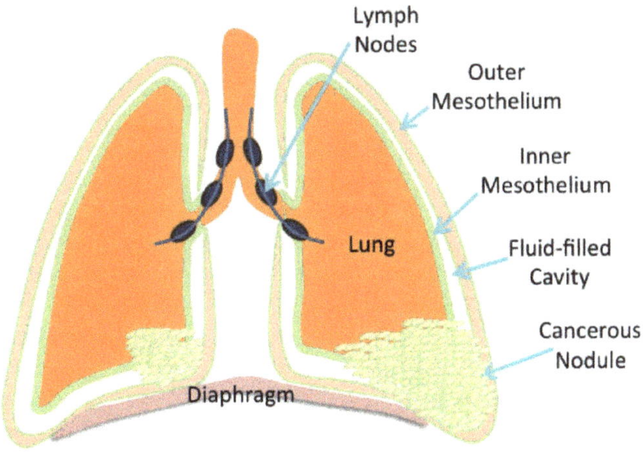

Abnormal growth around the lung.

To identify this growth, his doctors obtained a biopsy—a tiny bit of tissue from one of these clumps of growth. They then observed it under a microscope. Some of the cells in this biopsy appeared clearly abnormal. Their shapes and sizes suggested that they arose from *Siva*'s *mesothelium*, a membrane that covers his lungs and some other organs.

In case you are wondering what role the mesothelium might possibly play, just imagine how our lungs are cushioned from impact as we jump up and down, dive, fall, bang into each other, and so on. The mesothelium provides this cushion. It comprises two thin, protective linings, as shown in the picture above. The gap between these linings is filled with a few teaspoons of lubricating fluid, which provides the cushion.

Similar to the lung mesothelium, a pair of linings cushions the abdominal organs as well, as shown in the picture below.

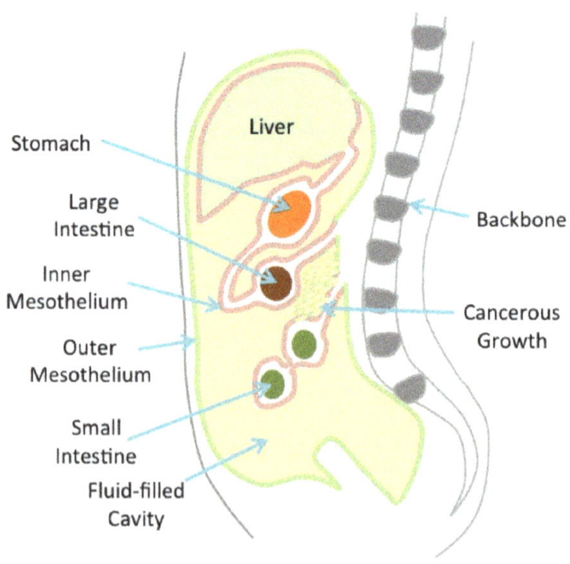

Abnormal growth in the abdominal mesothelium, in profile.

The gap between the inner and outer linings is much larger here, but, as before, it is filled with cushioning fluid. Presumably, some cancer cells from his lung mesothelium had broken off and lodged themselves here. These cancerous cells were secreting fluid, causing excessive accumulation in the gap between the two linings. This was the cause of *Siva*'s abdominal pain.

Siva was accordingly diagnosed with *mesothelioma*, a cancer of cells of the mesothelium. In due course, excess fluid could build up in his lungs as well, making it hard to breathe. Left to itself, his cancer could continue to grow and spread, invading organs, and proving fatal eventually. Indeed, statistics indicate that only 40% of mesothelioma patients survive a year past diagnosis, and only 10% survive five years past diagnosis.

There was no history of the disease in *Siva*'s family, or anything else to suggest an inherited cause. Then what caused cells in *Siva*'s mesothelium to lose their usual discipline and start dividing uncontrollably? And what tools could modern science and medicine offer to *Siva* to fight this cancer?

Asbestos?

The word *asbestos* has its origins in a Greek word which means "inextinguishable". For good reason—sheets and clothes made from this naturally occurring mineral have long been known to be heat and fire resistant. Roman legend mentions an emperor who would throw his asbestos-derived tablecloth into the fire at the end of a meal and marvel as it emerged unscathed. Over the years, this "miracle material" became more and more ubiquitous and found its way into all types of construction material, like pipes, cement, plaster, roofing sheets, and tiles, for heat and fire insulation. Soon enough, our buildings came to carry substantial amounts of asbestos.

As early as 2000 years ago, Pliny the Elder, the well-known Roman philosopher, noted that slaves working in asbestos mines suffered from sickness of the lungs and died at a relatively early age. This observation was reaffirmed again in the latter half of the

20th century. The aftermath of the second world war saw a massive emphasis on rebuilding and construction, employing workers in large numbers. These workers experienced prolonged exposure to asbestos. Over the next few decades, several of them developed mesothelioma.

We now know that asbestos used in construction material leaks microscopic fibers which we can inhale.[130] These fibers eventually make their way into the lung mesothelium and lodge themselves there. Here, they cause the production of *reactive oxygen species*, toxic molecules which react with genomic characters and cause various forms of damage, as we saw in the last story dealing with *Toto*'s Fanconi anemia. Sometimes, this culminates in cancer. Indeed, one study estimated that 1 out of 17 British men born in the 1940s and employed in carpentry for more than 10 years before the age of 30 would develop mesothelioma.[131]

The need to control the use of asbestos became increasingly obvious in the latter half of the 20th century for this reason. In a series of legislations in the 1970s and 1980s, the Environmental Protection Agency in the United States progressively banned several uses of asbestos. A final ruling was made in 1989 to ban almost all uses of asbestos. However, this ruling was appealed by the asbestos industry and overturned in a landmark judgement in 1991. So, several but not all uses of asbestos continue to be banned in the United States. The European Union completely banned all uses of asbestos in 1999. In contrast, the use of asbestos in India and China not only continues unhindered, but is growing rapidly, exposing large numbers of people to its deadly fibers.[132] It is hard to say, but given the ubiquity of asbestos in India, possibly, the genesis of *Siva*'s mesothelioma lay in some early over-exposure to asbestos fibers.

Virus?

Can cancer actually be transmitted from one person to another? Like the common cold, which spreads through the air? Or less easily, like AIDS, which requires contact of body fluids? Fortunately, the answer seems to be no—most cancers do not appear to be infectious.

However, observations made as early as 1842 suggested that certain cancers might be communicable. For instance, nuns in Verona seemed to develop cervical cancer much more rarely than married women.[133] Experiments performed in the early part of the 20th century then slowly started unraveling this infectious component of cancer.

In 1911, a farmer showed a hen with a large tumor (a *sarcoma*) in her breast to Peyton Rous, a pathologist at the Rockefeller Institute in New York. Rous' curiosity led him to explore whether the tumor could be transplanted to other chicken. To his surprise, it indeed could, though only to closely related chicken. Rous then filtered out hen cells and bacteria from the tumor and again tried the transplant. Rather surprisingly, this transplant too caused tumors in the recipient. Rous hypothesized that a virus might be carrying the cancer from one chicken to another.[134] Of course, microscopes of the day were not powerful enough to identify the virus. Further, electron microscopes that made viruses visible were only invented a few decades later.

We now know the virus in Rous' experiment, aptly named *RSV* or the *Rous Sarcoma Virus*. While most viruses destroy the cells that they infect, RSV alters the cell instead, making it divide more rapidly. Oddly enough, daughter cells resulting from this division too divide rapidly, even when there is no virus. How can that be?

It turns out that the virus, most cunningly, incorporates its own genes into the genomes of the cells that host it. One of these genes, called *v-src*, has now been identified as a gene responsible for inducing cancer.[135] A chicken cell cannot distinguish between its own genes and those which the virus introduces; it therefore executes recipes from both. The protein created from the *v-src* gene then triggers uncontrolled cell division. When this happens, the daughter cells too inherit this gene, and alongside, the disposition for uncontrolled division.

Of course, the Rous Sarcoma Virus involves chicken. What happens to us humans? Fortunately, this virus does not cause cancer in humans. But there are others. The first such virus was discovered in 1964 and is now named Epstein-Barr Virus after two of its three discoverers: Anthony Epstein and Yvonne Barr (the third was Bert Achong). Just a handful of other such viruses have been discovered since. And 10–15% of all human cancers are now understood to be caused by viruses.[136]

Of these, the greatest contribution comes from the *Human Papilloma Virus*, or *HPV*, which is responsible for 70% of the cervical cancer cases in the world. Remember nuns in Verona were observed to have a lower incidence of cervical cancer than married women; this is because the HPV virus spreads primarily through sexual contact. Genes provided by this virus waylay two important human genes that keep cell division in check. One of these is the *RB1* gene we saw in the story on *Ali*'s retinoblastoma. The other is a gene called *TP53*, arguably the most vocal gene in cancer. We will see why, soon.

In the early 1960s, a series of studies brought a new virus to the fore, in a most unexpected setting.[137] Vaccinations for polio were then made using rhesus monkey kidney cells. Some of these vaccines were found to harbor a new virus, called SV40. This virus usually kills the cells that it infects in the rhesus. Surprisingly, hamster cells did not die when inoculated; rather they started developing tumors. Millions of people were inadvertently exposed to SV40 via the polio vaccine between 1955 and 1963, in the United States and in Europe. This prompted fear of increased cancer rates among the vaccine recipients. Fortunately, several studies now suggest that this fear may be exaggerated.

We return now to *Siva*'s mesothelioma. In 1993, suspicion of SV40 as a cause for mesothelioma was spurred by the finding that Syrian hamsters when infected with SV40 developed mesothelioma.[138] The fact that traces of this virus were found in several mesothelioma patients[139] added to this suspicion. Not all studies have replicated this phenomenon though. And the mere presence of SV40 isn't proof that it is indeed the cause of cancerous transformation; that proof remains elusive. Human cells in a laboratory dish are indeed known to turn cancerous under the influence of SV40.[140] So, again, one cannot be certain, but it is possible that the SV40 virus played a role in the genesis of *Siva*'s mesothelioma.

Or Plain Random Chance?

What else could have caused *Siva*'s mesothelioma? The analysis of Knudson's observations from the story on *Ali*'s retinoblastoma suggested that randomly occurring somatic variants in the genome might be a cause (remember, somatic variants occur in our cells after we are conceived, and rarely pass on to our offspring). A very interesting observation, crystallized quite recently, reinforces this insight. This observation arose from a rather puzzling fact. Cells in the inner surface of the small intestine are exposed as much to pesticides and other chemicals in our food as are cells in the inner surface of the large intestine. Yet, tumors in the large intestine are much more common than those in the small intestine. Why so?

The picture below attempts to answer this question, using data from 31 different cell types (or *tissues*, as they are called).[141]

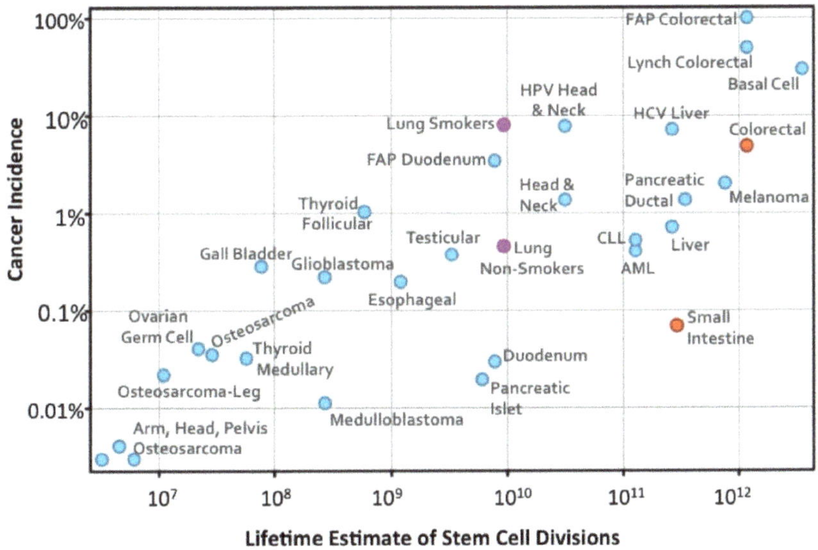

Cancer risk versus number of stem cell divisions in different tissues.[141]

If the answer suggested by this picture is not obvious to you, then read on. Recall stem cells from *Dia*'s vision loss story—cells in every tissue are lost continually, and stem cells divide continually to yield new cells that compensate for this loss. The point now is that the number of stem cells varies widely between tissues, as does the rate at which these cells divide. Taken together, the number of stem cell divisions over a lifetime varies widely from tissue to tissue. The spread of points in the picture above suggests that the risk for cancer is correlated to the number of stem cell divisions.

For instance, the lung sees about 10 billion stem cell divisions, while the large intestine, marked as colorectal in the picture, sees as many as 1000 billion. Accordingly, the lifetime risk of cancer is 0.45% for the lung, and roughly 10-fold higher at 4.8% for the large intestine.[141]

Of course, this correlation between stem cell division and cancer incidence is far from perfect, for the points in the above picture appear to spread out rather than stick to a tight straight line. Indeed, careful math shows that the number of stem cell divisions can explain, not all, but roughly 2/3rds of the variation in cancer incidence between tissues.

Interpreted in a way, a substantial part of our risk for cancer might just arise from the way we are made. Our cells have to divide billions to trillions of times over our lifespans. Each division has to copy the entire genome. Seemingly random errors creep in each time a copy is made. Over many cell divisions, these random errors accumulate, leading to cancer.

No doubt environmental influences add to this risk. As the above picture shows, the risk of lung cancer for a smoker is 18 times that for a non-smoker. In fact, the number is 23 times for males and 13 times for females, averaging to 18 on the whole.[141] Smoking leads to widespread genomic damage, so errors accumulate at a much faster rate with each cycle of cell division.

Regardless, the baseline risk of cancer is substantial even in the absence of smoking, and a majority of this risk appears to reside in factors well outside our control—random errors in the genomes of our cells over several trillions of cell divisions.

Of course, whether these random errors will cause cancer in any particular individual is hard to predict. Maybe the roll of the dice in this case was simply not in *Siva*'s favor. Whichever way, whether it was asbestos or a virus or just plain random chance or a combination of all of these, cancer had made its first move by announcing itself. It was now for *Siva* and his caregivers to respond with their moves.

Preparatory Moves

If Siva's tumor were localized, doctors could attempt to remove it surgically. They could remove just the lung mesothelium while sparing the lungs themselves, or they could even remove an entire lung. Unfortunately, neither was guaranteed to yield a cure. Microscopic traces of the tumor could be left behind and these could grow back into full-fledged tumors in due course. Radiation could potentially destroy these tumor traces. Even then, the expected survival after surgery would be only a few years.[142]

While *Siva* was young and healthy enough to tolerate both surgery and radiation, his cancer was not easily amenable to either. Some tumor cells had broken away from his primary tumor and spawned secondary tumors at several locations around his lungs. Worse yet, some tumor cells had moved to his *lymph nodes.*

The lymph system is a network of vessels within which circulates a colorless fluid, of course, called lymph. Cellular waste, microbes, and damaged cells all drain into the lymph. The junctions in this lymph network are called lymph nodes; here the immune system attacks and destroys invading microbes. Some vagrant tumor cells had broken away from their parent tumor and drained into the lymph system and thence travelled to these lymph nodes. Some had proceeded further and traveled all the way to the mesothelium of the abdomen and initiated new tumors there. Unfortunately, *Siva*'s cancer has spread quite a bit before it had announced itself.

Given this spread, approaches other than surgery and radiation were needed to restrain or eliminate his cancer cells. In this quest, his doctors faced a fundamental question: how could they distinguish between *Siva*'s cancer cells and his other normal cells, so they could focus their attack on these cancer cells alone. Indeed, what makes a cancer cell different from a normal cell?

Hallmarks of Cancer

The main hallmark that differentiates a cancer cell from a normal cell is that it divides much faster. To get to this state of fast division is not easy though, for nature poses several hurdles along the way. To overcome these hurdles, cancer cells must acquire several supporting hallmark traits.[143]

Let us start by asking the question: what induces a cell to divide? There must be an innate switch in nature, for our bodies are sculpted by carefully coordinating which cells divide and when they divide. A historic observation in 1950 by Rita Levi-Montalcini provided early glimpses into this mechanism.

Levi-Montalcini implanted a mouse tumor into a developing chick embryo. She was curious as to how chick cells would react to the presence of these foreign cancer cells. To her surprise, she noticed a particularly dramatic outgrowth of chick nerve fibers toward the implanted tumor cells. These nerve cells, although far away from the tumor, seemed to be responding to it by dividing fast. How could the implanted tumor cells induce the division of these nerve cells some distance away?

Today, we know that cells can release substances called *growth factors* into the bloodstream. These molecules are then detected by sensors embedded on the surfaces of other cells, possibly far away in the body. When growth factors attach to these sensors, a cascade of reactions is triggered within those cells, culminating in cell division, as shown in this picture.

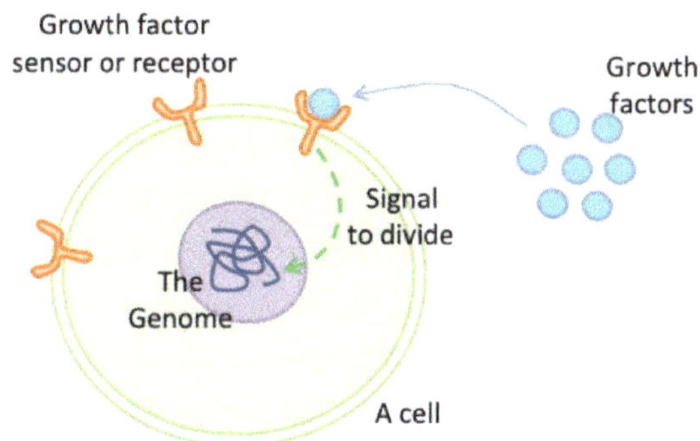

A growth factor induces cell division through its sensor, or receptor, as it is called.

Normally, growth factors are produced and released in carefully controlled ways so cells divide only when the need arises. Occasionally though, a cell can turn rogue by producing and releasing growth factors in excess, or insidiously switching on the reaction cascade in other ways even in the absence of growth factors. This is one of the key hallmarks of a cancer cell.

Then comes the next hurdle: so-called *tumor-suppressor genes* now block the way. For instance, remember the *RB1* gene and the *TP53* gene which the HPV virus inactivates? These genes, and some others, act like watchdogs, allowing cell division to proceed only when a number of normalcy conditions are met. The cell must inactivate one or more of these genes on its march toward turning cancerous. Indeed, we know that many cancer cells carry variants that truncate the *TP53* recipe prematurely.

Continuing along its obstacle course, the cell now confronts the programmed death machinery, which prefers to induce cell death when it senses that things are not going as per plan, rather than continue abnormally. The cell must defuse this machinery for it to turn cancerous.

Then, as the cell starts dividing faster, it needs nutritional supplies at a rate higher than what a normal cell would demand. For this, it must induce the creation of blood vessels around itself and also make changes to the way it generates energy to fuel its rapid progress. If it fails to do so, it risks running out of supplies and dying in the process.

At this point, the cell has all the wherewithal to divide rapidly. Its next roadblock is an innate limit on the number of generations of cell division that it can go through, typically just 40–60, before it can divide no further. At this point, it either settles into a state called *senescence* without any capability to divide further, or dies. To continue its march toward cancer, the cell must escape this innate limit and continue dividing.

After all this, the cell faces yet another adversary: the body's immune system. Typically meant for dealing with invading microbes rather than the body's own cells, the immune system detects fragments of proteins that appear foreign, and then destroys cells which carry these foreign-looking fragments. Cancer cells, in all the genomic trial and error needed to overcome the hurdles mentioned above, cannot often help but create several foreign-looking proteins, potentially exposing themselves to the wrath of the immune system. So, the cell must evolve ways by which it can evade the immune system before it can survive and thrive as a cancer cell.

Finally, the cell, now fully in cancerous mode, will run out of space where it is if it continues to divide rapidly. It must find a way to break away from the bonds that its neighbors and the surrounding matrix impose on it, and travel through the lymph network or the bloodstream to another part of the body, where it can establish itself afresh.

So many hurdles to cross. It is a wonder that cells manage to get past all of these hurdles. Of course, somatic variation in their genomes is what enables them to do so.

You might remember from *Toto*'s story on Fanconi anemia that somatic variation is induced into the genome continually in the background. Not all variants are helpful to a cell in its quest to turn cancerous. However, some variants do help. Cells with such variants survive preferentially and increase their numbers, enabling them to experiment with further variants and cross further obstacles. The pattern of somatic variants in the cancer cell's genome can therefore provide a glimpse into how it went about overcoming these obstacles. What did this pattern look like in *Siva*'s cancer cells?

The Cancer Genome

To identify the pattern of somatic variants in *Siva*'s cancer cells, we need a few of these cancer cells. Accessing these cancer cells is not as easy as drawing a bit of blood or saliva from a person. Nevertheless, one can access them using a routine biopsy procedure.

Usually, this procedure yields a mix of cancer cells and normal cells. Cancer cells then have to be carefully separated, often by a trained eye which can tease apart clusters of cancer cells within an ocean of normal cells. This separation is far from perfect, but good enough to proceed.

We then sequence the genomes in these cancer cells and the normal cells separately, much as we did in the previous stories. By spotting differences between the cancer genome and the normal genome, we get a catalog of the somatic variants in *Siva*'s genome. These are character changes which his cells acquired on their path to turning cancerous, well after he was conceived.

Before we investigate these variants, a brief digression on the picture that has emerged from a similar exercise performed on thousands of cancer patients. Of the 20,000 or so genes in the human genome, a typical cancer genome appears to have somatic variants which cause recipe changes in relatively few genes: just a few tens to a few hundreds.[144] As expected, lung cancer cells from smokers occupy the upper end of this spectrum. So do skin cancer cells,[144] on account of exposure to ultraviolet rays.

These recipe changes include all types of changes that we have seen in the stories in this book. In addition, cancer genomes seem to carry some more drastic genomic changes—gene *amplifications*, where many copies of a gene are created, possibly causing it to function in overdrive, or conversely, gene *deletions*, where entire copies of a gene are deleted.

A striking fact here—the collection of recipe changes varies from one cancer patient to another. In that sense, one person's cancer is different from that of another. Each person's cancer cells appear to have navigated slightly different routes to obtain the hallmarks discussed earlier. Cancer, you could therefore say, is not one disease, but many diseases.

Another striking fact as one observes recipe changes in the tumors of many many cancer patients—some genes, counting to 140 or so, seem to appear again and again.[144] These so-called *driver genes* probably play serious enablement or disablement roles in making a cell cancerous, and successful cancer cells often show evidence of having tampered with their recipes so that enablement is enhanced and disablement is reduced.

Even among these driver genes, a few seem to dominate the landscape. For instance, considering hundreds of tumors whose genomes we have sequenced at Strand and found variants in, more than 80% carry recipe changes in one of just four genes: *TP53*, *PTEN*, *PIK3CA* and *KRAS*. Even more surprisingly, as many as 50% of these tumors carry recipe changes in just one gene: *TP53*. 50%! There are an estimated million new cases of cancer in India every year. A similar number is true for the United States. And half of these are likely to have recipe-altering somatic variants in *TP53*. This gene is clearly a key sentinel on the path from normalcy to cancer. Hence the name *TP* for *tumor protein* (the number 53 denotes its weight).

Which were the genes driving *Siva*'s cancer? Any given tumor specimen has recipe-altering somatic variants in only a handful of these 140 or so driver genes. To find the answer for *Siva*'s cancer, we would have liked to sequence the exons of all these 140 or so driver genes. However, not all of these genes can help *Siva*'s doctors in planning their next move. So, for cost reasons, we sequence key exons from just 48 carefully chosen driver genes. And we find three genes with key recipe-altering somatic variants.

Not surprisingly, two of these genes are among those we just mentioned: *TP53* and *PTEN*. The third gene is also quite commonly found to carry somatic variants in cancer genomes—a gene called *EGFR*.

What do these genes tell us about *Siva*'s cancer? And how could his doctors use this information in their fight against the cancer?

Where to Attack?

In the middle of the 20th century, it was yet unknown if a drug could tame cancer by providing long lasting remission in a sizable fraction of cancer patients, let alone a complete cure. So, if surgery failed or was infeasible, as in *Siva*'s case, then all hope was given up. That is no longer the case today.

Doctors today have several drugs available to treat cancer. These drugs take one of two broad approaches—either they attack the main hallmark of a cancer cell, which is rapid cell division, or they attack one or more of the several supporting hallmarks we saw earlier.

Impeding rapid cell division with chemicals has been the more dominant approach for several decades. The origin of some of these so-called *chemotherapies* dates back to World War II when a chemical called nitrogen mustard was stockpiled as a chemical weapon. Accidental release of this chemical in an air raid incident affected several people in 1942. Medical examination of the survivors showed much lower than normal counts of *lymphocytes*, a type of white blood cell. This hinted that division of stem cells in our bone marrow, where these lymphocytes are generated, was suppressed by nitrogen mustard. With further research, it became the first chemotherapy to be approved by the United States Food and Drug Administration (FDA) in 1949 for treating cancer of the lymphocytes. Since then, several more chemotherapy drugs have been discovered.

Of course, the challenge with these chemotherapy drugs is that they are blunt instruments. They attack all rapidly dividing cells, cancerous or not. Stem cells in our bone marrow also divide, maybe not as fast as cancer cells do, but nevertheless fast enough to come under attack. The cells that line our gut as well as our hair follicle cells too divide often. Chemotherapy attacks all these cells as well. The consequences can be very unpleasant—low blood counts, poor immunity, nausea, diarrhea and hair loss.

The alternative for *Siva*'s doctors was to understand precisely how his cancer cells had obtained their supporting hallmarks, and then use drugs to interfere with these precise mechanisms. These so-called *targeted therapies* come in a variety of flavors, depending on the hallmark being attacked.

For instance, remember cells grow and divide when a growth factor molecule attaches to a sensor sticking out of the cell. This attachment triggers a cascade of reactions within the cell, culminating in division. If a cell cleverly switches on this reaction cascade, say by overproducing and releasing growth factors, it could turn cancerous. A targeted drug molecule that gingerly blocks out the sensor's access to the growth factor without switching on the reaction cascade could well do the trick in this case. Hopefully, such targeted therapy attacks cancer cells while sparing most other cells, or at least those that do not rely on this particular sensor for their normal course of division.

Another example: a tumor cannot grow large unless it increases blood supply to itself by spawning the creation of new blood vessels. A growth factor molecule called *VEGF* plays a key role here. Fortunately, while *VEGF* is instrumental in spawning new blood vessels to feed the tumor, its role in normal blood vessels is more subdued in adults. So, a targeted drug that defuses *VEGF* helps in containing cancers while keeping side effects to a manageable level.

Yet another example takes us back to *Toto*'s case in the previous story. Recall how our cells use *homology* search to repair breaks that occur in both strands of the genome—missing characters are filled in by reading off another genomic copy, thus repairing the break. In some tumors, this repair mechanism is compromised by genomic variants. Drugs which promote the creation of double-stranded breaks in such tumors then accelerate the rather haphazard jumbling of chromosomes we saw in *Toto*. The chaos this wreaks on the genome pushes tumor cells toward programmed death, thus containing the cancer.

Another particularly interesting example comes from how a cancer cell evades the immune system. The so-called *T-cells* of the immune system are meant to recognize intruder cells and kill them. They do so by *feeling* their quarry using multiple sensors. One sensor looks for specific molecules sticking out of the target cell. Only if it finds these molecules does it launch an attack on its quarry. Even then, it waits, as other sensors interact with other proteins sticking out of the target cell. Some of these proteins encourage the T-cell to go ahead with its attack while others discourage it, possibly a balancing mechanism to ensure that immune attacks are launched with much discretion and measure. Coming back to cancer cells: they do sometimes present molecules on their surface which T-cells need for launching their attack. However, cancer cells are also clever enough to put out other discouraging proteins which keep the T-cell at bay. By using a drug that blocks the relevant sensors, this discouragement can be removed, thus inducing the T-cell into attacking its quarry.

Over the last two decades, an increasing number of these targeted therapies and immunotherapies have been approved for treatment by the United States FDA. So, *Siva*'s doctors had at their disposal an arsenal of chemotherapies, targeted therapies and immunotherapies to deal with his cancer. Yet, they were not spoilt for choice, for several reasons.

A key reason is that a drug that works on, say, breast cancer, need not have any effect on, say, lung cancer. Each cancer seems to respond in its own way. And even within breast cancers and lung cancers, there are many sub-types, with some responding to a particular therapy and others not. This poses a tremendous challenge: for each drug, clinical trials have to be conducted in each type and sub-type of cancer, to prove that the drug indeed works before it takes its place in mainstream medical practice. There are tens of types and sub-types, making this an expensive and time-consuming affair. As new drugs are discovered continually, one might even think of this process as endless. The net result is that only a small number of drugs are applicable for treating cancer of any given type.

Added to this is the issue of economics. Clinical trials are expensive, so pharmaceutical companies often focus on more common cancers, which yield a larger market. Rarer cancers take a backseat sometimes. So, the choices available for a rarer cancer type, such as *Siva*'s mesothelioma, are considerably lower than for other more common cancers of the lung. In addition, the huge costs of discovering a drug and taking it through clinical trials make cutting-edge cancer drugs very expensive, particularly in countries like India.

All this means that *Siva*'s doctors had limited options to treat his cancer. The National Comprehensive Cancer Network (NCCN), an alliance of 26 of the world's leading cancer centers, captures the outcomes of various clinical trials into a collection of guidelines for treating and managing cancer. Each type of cancer has its own dedicated set of guidelines. These guidelines are widely used by doctors. Given the variability of drug response from one cancer type to another, doctors rarely stray from these guidelines, at least in their initial course of treatment.

Of course, there are guidelines specifically for lung mesothelioma as well.[145] Unfortunately, these guidelines only cover blunt attacks with chemotherapies, for no targeted therapy or immunotherapy has been sufficiently proven in large clinical trials yet for mesothelioma. So, *Siva*'s doctors had to make their first few moves using these blunt attacks alone. The fact that his cancer was driven, at least in part, by the *TP53*, *PTEN* and *EGFR* genes, was not of relevance in their treatment. At least, not yet.

The Opening Move

The first move made by *Siva*'s doctors was aimed at interfering with the rapid cell division of his cancer cells. Remember, an entire copy of the genome is made during cell division. This process requires the core building blocks of the genome, the characters A, C, G and T, in sufficient quantities. These four molecules are manufactured inside our cells. Drugs which interfere with this manufacturing process can make these building blocks scarce, thus slowing down cell division.

Indeed, such a drug, called *pemetrexed*, is recommended by the NCCN as part of its guidelines for treating mesothelioma.[145]

There are other ways as well to interfere with the cell division process. One of these again goes back to *Toto*, the boy with Fanconi anemia in the previous story, whose ability to resolve *crosslinks* was highly compromised. Remember crosslinks—those unauthorized connections which effectively block cell division, making genomic repair genes work hard? This picture should refresh your memory.

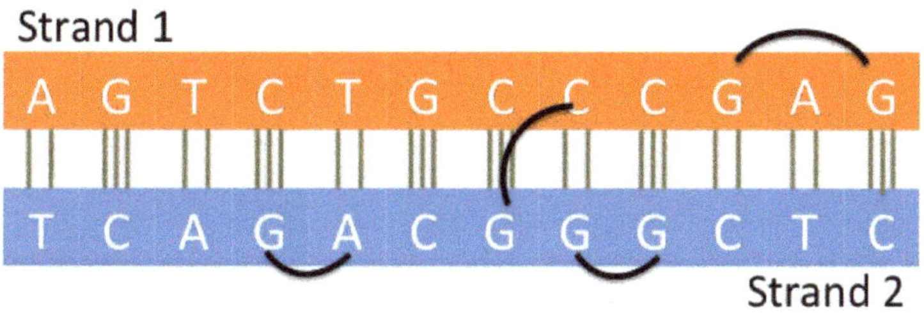

Crosslinks caused by cisplatin.

Some drugs, like *cisplatin* and *carboplatin*, can indeed introduce crosslinks in large numbers in the genome. As their names indicate, these drugs are compounds of platinum, which is among the most expensive precious metals on earth, ahead of gold. A combination of the drugs pemetrexed and cisplatin is commonly used as the first line of treatment for mesothelioma.[146] This was the combination doctors prescribed for *Siva* as their first move.

Note a point of irony here. The boy with Fanconi anemia had much-reduced ability to resolve crosslinks. Not crosslinks artificially induced by drugs, but simply those that occurred in the normal course of life. This made his risk for cancer over three decades or so very high. And here was *Siva*, a cancer patient, on whom standard treatment forced a barrage of crosslinks at a much higher than normal rate, in an attempt to cure his cancer.

Of course, the goal of inducing crosslinks in large numbers was to challenge the genomic repair machinery beyond its capability. Cells would then detect that the genome is going unrepaired and invoke programmed death. So, in the short term, introducing crosslinks would kill dividing cells. In the long term, the risk of cancer would increase. For *Siva*, the battle was clearly focused on the short term, for the average life expectancy of a mesothelioma patient is only a few years. The mission at hand was to kill cancer cells even at the expense of an increased long-term risk for cancer.

Indeed, several of *Siva*'s cancer cells died under the onslaught of pemetrexed and cisplatin. A CT scan showed that some tumor nodules had become smaller and others had vanished. And *Siva*'s abdomen pain reduced substantially. The battle had been won.

However, the war wasn't over yet. Some cancer cells survived and continued to experiment with various genomic combinations, via genomic surgeries performed during the genome repair process. The occasional cell then evolved a notorious combination, possibly just by chance. It managed to shut its doors to cisplatin (or at least so we are guessing), which no longer could make its way into the cell and create crosslinks as it normally would.[147] It also managed to suppress the innate urge for programmed death when things go wrong in the genome. This cell became resistant to cisplatin, and continued to divide recklessly. A year and a half later, as more such cells accumulated, the cancer was back.

More Blunt Attacks

Siva's cancer cells were now resistant to the pemetrexed/cisplatin combination. So, his doctors needed to switch strategy. Their next move, in accordance with the guidelines provided by the NCCN,[145] was to replace cisplatin by its cousin, a drug called *carboplatin*.[148]

Carboplatin, like cisplatin, works by inducing crosslinks in the genome, albeit less intensively and with far fewer side effects.[149] The attack by pemetrexed and carboplatin pushed *Siva*'s cancer to the back-foot once again. The tumors started to shrink and the symptoms abated.

The cancer again lay low for a while, figuring out its next move.

But move it did. After about nine months, it was back. The tumors were growing once again and there was further growth in the abdomen.

It was now close to three years since *Siva* had been diagnosed. Already longer than the typical survival period from diagnosis for a mesothelioma patient. Possibly, *Siva*'s relatively young age of 37 at diagnosis had helped. *Siva*'s doctors now needed to make their next move to keep him alive and relatively healthy. They added a drug called *gemcitabine* to the pemetrexed/carboplatin combination, again in accordance with treatment guidelines provided by the NCCN.[150]

Gemcitabine has its own clever way of stalling cell division.[151] Remember, cell division first requires that an entire copy of the genome be made. This is done by assembling the copy one character at a time. Of course, this requires an ample supply of genomic characters. Gemcitabine depletes this supply. It then posits as a trojan, offering itself (more precisely, a modification of itself) as a replacement character. The genome copying process falls for this trojan trap: it uses this replacement character in the copy being assembled. Only after it has added this and then one more character to the copy does it realize that this replacement character is a trojan that allows no further addition of characters. The genome copying process stalls, again prompting the cell to undergo programmed death.

Fortunately, the cancer in the lung responded to this new regimen. Unfortunately, the cancer in the abdomen remained stubborn, and continued to grow. Presumably, these cells had worked out a genomic combination that defused all the drugs tried so far. Close to four years from diagnosis, *Siva*'s doctors were running out of options. The cancer was threatening to move toward its endgame.

The only other drug which finds mention in the treatment guidelines provided by the NCCN, the official standard of care, was a drug called *vinorelbine*.[145] This drug too blocks cell division in its own distinctive way. Unlike the earlier drugs mentioned above, Vinorelbine does not obstruct genome copying itself. Instead, it obstructs the creation and function of a scaffold structure which supports the copying of the genome.

So, *Siva*'s doctors tried combinations of pemetrexed, carboplatin and vinorelbine. Some caused serious side effects and had to be pulled back. None seemed to unsettle the cancer, which, by now, had honed its defenses, and continued its unrelenting march forward.

The NCCN guidelines had nothing more to offer *Siva*. All the blunt attacks prescribed there no longer challenged his cancer. *Siva*'s doctors now needed to think out of the box to make their next move. They had to venture beyond established guidelines based on the outcomes of large clinical trials, and step into the world of experimentation.

Remember our sequencing of the genome from *Siva*'s cancer cells? These cells had cleverly modified their genomes with recipe-altering variants in the *TP53*, *PTEN* and *EGFR* genes. Could this information help suggest the next move to *Siva*'s doctors? A tour of each of the three genes in turn might help answer this question.

The Guardian Lowers its Guard

We start with *TP53* to see if it might yield a clue. This gene carries the sobriquet *guardian of the genome.*[152] When cells detect genomic damage, they activate the so-called p53 protein derived from this gene, which promptly puts a brake on cell division. This gives time for the genome to be repaired. And if the genome is determined to be beyond repair, p53 initiates programmed death, ensuring that cells with damaged genomes do not replicate. Similarly, when genes which trigger cell division go on an overdrive, p53 provides a protective umbrella by inducing programmed death.

Recently, an interesting anecdote has emerged around this gene. Remember a broad trend that we saw earlier in this story—more the number of cell divisions, higher the risk of cancer. One would think therefore that large, long-lived animals have a high risk of cancer. Elephants are very large and live for several decades. Yet their mortality from cancer appears to be a fraction of ours, a fact referred to as Peto's paradox. It turns out that while we have only two versions of *TP53* in our genome, elephants have at least 40 versions![153] Possibly, this accentuates the urge to commit programmed death in response to genomic damage, thus reducing cancer risk.

Given how important *TP53* is, it is no wonder that many tumors appear to have defused it along the way. *Siva*'s cancer too had done so, using two recipe-altering variants which were apparent when we sequenced the genome in his cancer cells. We don't know if both variants were present in the same cancer cell, or whether some cancer cells had acquired one variant and some others had acquired the other. Indeed, different cells from the same cancer have been observed to carry different sets of variants. We do know that neither variant was present in his normal cells.

Both variants were missense variants that caused only subtle recipe modifications in the *TP53* gene. It is very likely that many tumor cells had only one version of the gene so modified; the other was probably untouched. One would think that this good version would compensate for the mutated one, thus keeping the cancer in check. Unfortunately, this wasn't the case. The reason needs some explanation.

The p53 protein has two key parts, colored distinctively in the picture below. The bottom part is used to attach to the genome, while the top part enables four p53 proteins to tether to each other to form a quartet.

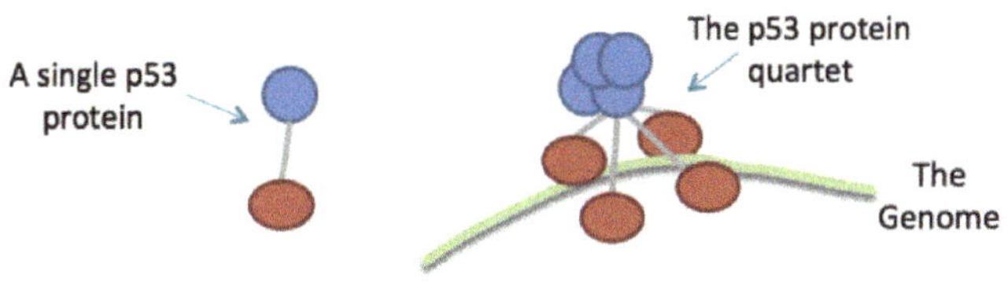

The p53 quartet attaches to the genome, turning nearby genes on or off.

When this quartet attaches to a specific place in the genome, a nearby gene is turned on (or off); which means, more (or less) recipe execution on that gene. Thus, by carefully attaching to specific places in the genome, this quartet is able to turn on or off specific genes needed to accomplish the intended agenda—stalling cell division, and inducing programmed death, if needed.

To attach to all the required places in the genome, the four members of the quartet must form bonds between their bottom portions.[154] This allows for a better grip on the genome. When this bonding is disturbed for some reason, the quartet can attach to only a fraction of the places it ought to. Genes that are then turned on or off do stall cell division in response to genome damage but do not induce programmed death.[154] One defense against cancer is then lost.

Both the variants in *Siva*'s cancer cells affected the bottom portions of the p53 protein, thus compromising bonds between quartet members modestly.[155] However, since these variants likely appeared in only one of the two *TP53* gene versions, not all the quartets felt their impact. Quartets in which all members came from the intact version continued to function as expected. Quartets with one mutated member also managed to retain much of their activity. Only quartets with two or more mutated members could no longer attach with sufficient strength to the relevant genomic locations.[156] Unfortunately, at least 68% of the quartets in *Siva*'s cancer cells fell into this category (you can do the calculations yourself if you like—for every sixteen quartets, one has no mutated members, four have one mutated member, six have two, four have three, and one has all four members mutated). The net result was that induction of programmed death no longer happened as strongly as it should, thus clearing the way and allowing *Siva*'s cancer to progress.

Was there a drug that could somehow neutralize these mutated p53 protein copies, leaving the good ones to do their job? That would resurrect programmed death, and possibly contain the cancer.

Unfortunately, the combined efforts of several pharmaceutical companies and academic researchers haven't produced such a drug yet, though research is ongoing on several candidates.[156] Given that 50% of cancers have mutated *TP53*, such a drug, as and when it arrives, could be in great demand. For now, we move on to the next gene we found mutated in *Siva*'s cancer and see if it offers some help to his doctors in planning their next move.

A Sensor on Overdrive

Recall Levi-Montalcini's rather surprising observation when she implanted tumor cells into chick embryos and found nerve cells some distance away dividing. This happened because tumor cells secreted growth factors into the blood. Subsequently, sensors on the surface of nerve cells detected the presence of these growth factors and initiated cell division in response.

A sensor, or receptor as it is called, often straddles the cell boundary, sitting partly outside and partly inside. When a growth factor attaches to its outside, something changes on its inside as well, either its shape or some marks on its amino acids (similar to the marks on the *RB1* gene in *Ali*'s retinoblastoma story). This sets off a cascade of reactions inside. These reactions then carry the signal of the growth factor's arrival to the genome, eventually resulting in cell division.

Now imagine a cell with receptors on its surface. These receptors must wait for the arrival of the growth factor before they initiate the reaction cascade that leads to cell division. Occasionally, something strange happens. A recipe-altering character modification makes the receptors jump the gun and initiate the reaction cascade all by themselves, without waiting for the arrival of the growth factor. Surprising as it may be, this appeared to be the case for *Siva*, the receptor in question being *EGFR*, short for *epidermal growth factor receptor*.

EGFR senses a few different growth factors. When these growth factors attach, two *EGFR* receptors come together to form a pair, as you can see in this picture.

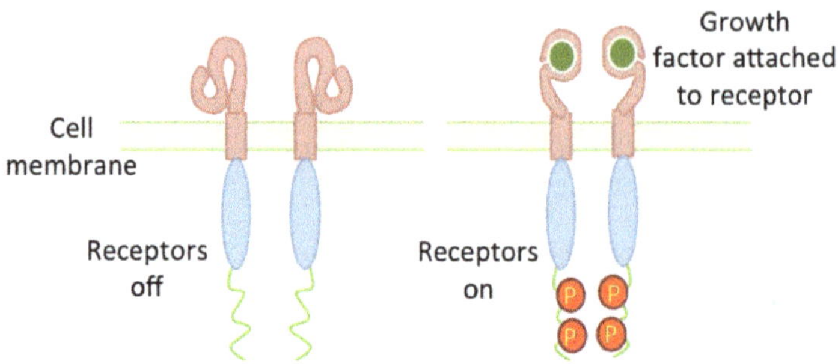

A pair of EGFR receptors. On the right, the marking that turns them on when the growth factor attaches.

A further change is needed before the two receptors can initiate the reaction cascade needed for cell division. This change, analogous to turning a switch on, requires the addition of special marks to each member of the pair. These marks are the same phosphoryl groups we saw in *Ali*'s retinoblastoma story, depicted by P in the picture above. In a reciprocal act, reminiscent of the mutual garlanding or ring exchange seen in weddings, each member of the pair puts marks on the other. The switches are on now, and the project to divide is initiated.

This usual mechanism had been hijacked somewhat in *Siva*'s cancer cells. A single character modification in some of his cancer cells caused the 725th amino acid of *EGFR*, usually a *Threonine*, to become a *Methionine*. This made the pair so over-eager that they jumped the gun to mark one another to an extent even when no growth factor was attached.[157] The downstream cascade inducing cell division was firing away consequently, even in the absence of the growth factor.

This excessive firing by *EGFR* was one of the likely hallmarks of *Siva*'s cancer. p53 would usually induce programmed death in such cases, thus protecting cells from excessive cell division.[152] But the recipe changes in *TP53* we saw earlier had lowered this guard at least a bit. The combination then provided *Siva*'s cancer an unhindered path to progress. What move could *Siva*'s doctors make to control this progress by suppressing *EGFR*'s autonomous, unprovoked firing?

A Sharper Attack

By now, *Siva*'s cancer had gathered much momentum, having evolved mechanisms to sidestep attacks with the likes of cisplatin and carboplatin. *Siva*'s doctors had to attack where the cancer still hadn't built up its defenses. They had to find a way to stop *EGFR* from firing indiscriminately. For this, they turned to a drug called *erlotinib*.

Erlotinib's expertise lies in blocking the part of *EGFR* that marks its mate. With the marking process becoming more sluggish, the downstream reaction cascade fires less often, thus arresting cell division on account of misfiring by *EGFR*. This could potentially rein in *Siva*'s cancer. Actual verification, of course, requires clinical trials. And several such trials have been conducted.

Trials on patients with lung cancer who had developed resistance to the likes of cisplatin and carboplatin have shown that about 7–9% of these patients do respond to erlotinib.[158] The patients in these trials were chosen more broadly and probably comprised a mix of those whose cancers were driven by *EGFR* and those whose cancers weren't. One would expect that patients with recipe-altering variants in *EGFR* that make it over-eager to mark its mate would respond even better. Indeed, the response rate for these patients was much higher, at 27%.[159] Even this may appear small to the uninitiated, but for those who have seen cancer from near, every bit counts.

Based on the above trials, the United States FDA approved in 2004 the use of erlotinib for patients with lung cancer who had developed resistance to chemotherapy. More recently, in 2013, the FDA approved erlotinib as the very first line of therapy for lung cancer patients with a few specific recipe-altering variants in *EGFR*.

Siva's cancer, a mesothelioma which arises not in the lungs but in the linings surrounding the lungs, wasn't quite lung cancer of the type for which erlotinib was approved. Therein lies an interesting dilemma: can drugs proven to work for one type of cancer work for another?

Strictly speaking, a lung cell and a mesothelial cell turn different sets of genes on and off; so they could have different responses to the same therapy. Traditional wisdom therefore doesn't easily lend itself to transferring therapies from one cell type to another. Unfortunately, the use of erlotinib in mesothelioma itself is far from proven, with some trials showing little effect[160] and others still ongoing.[161]

The other side of the argument is no less compelling, even if hard to prove. *Siva*'s cancer cells were indeed driven by a variant in *EGFR*, roughly of the type erlotinib is effective against. And given *Siva* had run out of all standard options, his doctors had to try an experimental option. Erlotinib was the natural option to try.

Siva's cancer responded well to this new move. The tumors started reducing or at least stopped growing and his symptoms stayed in control without many side effects. The cancer appeared to be have been reined in.

But, yet again, about nine months later, the cancer figured out a way to get around the treatment and started growing again. Yet again, *Siva*'s doctors needed to understand how the cancer had circumvented erlotinib, and decide their next move accordingly. Could the third gene which carried a recipe-altering variant in *Siva*'s cancer cells provide further options?

Another Brake Removed?

Remember, growth factor attachment to *EFGR* sets off a cascade of reactions inside the cell, inducing it to divide. In fact, it sets off not one, but several, cascades. One of these cascades is of particular interest to this story and appears below, albeit in much simplified form.

Each step in this cascade involves one protein influencing another—by either stepping on the brake to stop a particular function, or stepping on a gas pedal to activate that function. The colors or arrowheads should tell you which is which.

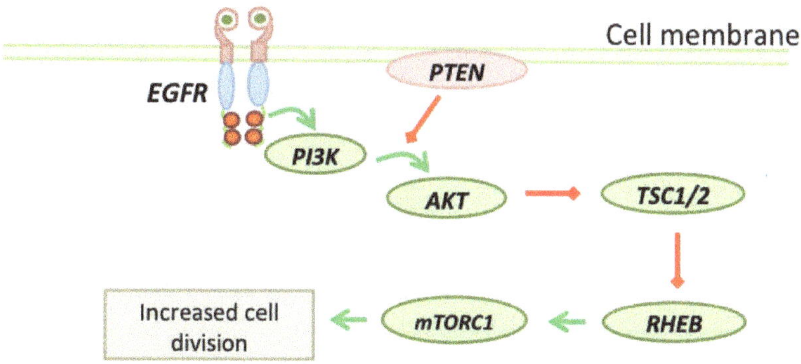

A reaction cascade initiated by EGFR. Green arrows with larger arrowheads indicate activation and red arrows with short arrowheads indicate brakes.

Now, follow the cascade from *EGFR* to *mTORC1*. There are two brakes that appear along the way. Of course, a brake applied on a brake is tantamount to stepping on the gas pedal. So, *Siva*'s cancer cells, which had acquired a variant in *EGFR* making it fire spontaneously, had managed to activate *mTORC1* through this cascade. In turn, *mTORC1* had pushed his cancer cells to divide faster.[162]

Fortunately, cells have additional control on this cascade in the form of a brake applied by *PTEN*. If this gene doesn't ring a bell, then recall it was the third gene in which we found a recipe-altering variant in *Siva*'s cancer cells.

Siva's doctors had attempted to stop *EGFR* from firing spontaneously using erlotinib. For a while, it appeared as if they had succeeded in their task. Then, maybe, some cancer cells evolved further variants in *EGFR*, enabling it to get around the erlotinib blockade. Consequently, *EGFR* continued to fire in these cells. Normally, *PTEN* would have checked the effects of this misfiring by keeping the cascade under control. In *Siva*'s cancer cells, variants in *PTEN* had possibly nullified that brake as well.

Our study of *Siva*'s genome showed two such variants: one in amino acid 233 and another in amino acid 247. Both these variants are exceedingly rare. One has been seen only in a single cancer patient before. The other has never been reported before. Much further experimentation will be needed to establish if these were indeed instrumental in removing the brake. Of course, *Siva*'s doctors could not wait that long to make the next move.

The cancer was pushing ahead and demanded a more immediate answer. Research on other variants located roughly in the same region of *PTEN* as *Siva*'s variants showed that 44% of these do indeed compromise the brake.[163] Given that *Siva* was running out of options, assuming that the *PTEN* brake was indeed removed was a risk worth taking. Even so, could *Siva*'s doctors use this information to make their next move?

The Next Move

A brief aside—a drug called *everolimus* was being tested as part of a clinical trial on 37 bladder cancer patients. The trial itself failed because the drug did not produce a significant effect. There was one exception though—a patient who showed a dramatic response. This begged the question: what was special about this one patient? What made everolimus work in this patient but fail in many others?

Sequencing the cancer genome of this patient gave the answer—a variant in the *TSC1* gene[164] that truncated its recipe prematurely. If you look at the cascade in the picture above, you will notice that *TSC1* keeps a brake on *mTORC1*. With *TSC1* gone or at least reduced, the foot was partly or fully off the brake, and *mTORC1* was firing more freely than usual in the patient in question, inducing rapid cell division. Everolimus, the drug being tested, seems to have put the brake back on *mTORC1*, as it was intended to do.

In a nutshell, the patient in question responded because everolimus compensated for the genomic error that drove the cancer in the first place. Other patients responded less favorably because the gene variants that drove their cancers were probably different.

Coming back to *Siva*, variants in his cancer cells could have possibly taken the *PTEN* brake off, causing *mTORC1* to fire. Everolimus could potentially put that brake back on *mTORC1*. For *Siva*'s doctors, this was a natural candidate for their next move.

Studies had shown that the side effects of combining everolimus with erlotinib were manageable.[165] Accordingly, and after appropriate permissions for trying experimental therapy, *Siva*, was put on that combination. Erlotinib would restrict *EGFR* from firing indiscriminately, and everolimus would apply an additional brake on *mTORC1*.

Siva had cycled through several therapies earlier. Each had pushed the cancer back and bought *Siva* time. But his cancer was now resistant to all these therapies. Would the combination of everolimus with erlotinib give *Siva* a fresh new lease of life?

Wrapping Up

The genome in our cells is constantly under attack: from various chemicals we inhale or ingest, from sunlight and other forms of radiation, and from viruses which infect us. Added to these are genome-copying errors which occur when our genome is copied trillions of times to create new cells. These cause somatic variants—variants which we do not inherit from our parents. The accumulation of sufficiently many such variants in a single normal cell sometimes transforms that cell to a cancer cell.

This evolution often takes years. Typically, half or more of the somatic variants which appear in a cancer cell occur before the cancer begins.[144] These are not the main drivers of the cancer, so they just ride along without tipping the cell over into a cancerous state. Then the first driver variant occurs, initiating the formation of a slow-growing tumor. Cells now divide slightly faster, often only slightly faster, than the rate at which they die. Over many years, this slight imbalance is sufficient to produce a large pool of cells. One of these then receives a second driver variant, which confers an even faster division rate to this cell and its descendants. This snowballing effect eventually results in a few cells accumulating several driver variants, leading to a fast growing, malignant tumor.

By the time such tumors announce themselves, they are often too diffused to eliminate completely by surgery and radiation. The only possible treatment then is with drugs that can travel to all parts of the body in the bloodstream. Some such drugs, like cisplatin and carboplatin, have blunt effects, broadly killing all dividing cells. Others like erlotinib are more specific and target only cells which have substantial amount of *EGFR*. Knowing which variants are driving the cancer can help pick the right drug for a given patient.

But cancer is a dynamic adversary, reacting and evolving with every such drug. Several cancer cells do die with each attack providing some respite. Sometimes, this brings a happy end to the ordeal. Very often though, the saga continues. Cancer cells with their rapid rate of division and much reduced levels of genome repair are able to experiment rapidly with new genomic characters and evolve further somatic variants which make them resistant to these drugs. The disease then comes back, with renewed vigor. A different drug has to be tried then.

Indeed, several such moves had been tried by *Siva*'s doctors, with each such move buying a fresh lease of life, albeit temporary. Each time the cancer had made its countermove and bounced back. One by one, possible options for treatment had been eliminated, leaving *Siva*'s doctors struggling for new options. A peek into the genomes of *Siva*'s cancer cells suggested first erlotinib, and now the erlotinib–everolimus combination.

The erlotinib–everolimus combination pushed *Siva*'s cancer to the backfoot once again, giving him yet another fresh lease of life. A fresh lease of modestly good quality life in fact, for his symptoms and side effects were not too debilitating.

Of course, a complete cure would have been a miracle. Typically, the only cure one expects with cancer is when it is localized and can be removed surgically. Once it spreads, it is generally considered incurable. The battle becomes one of buying time—good quality time, in particular. For *Siva*, the response lasted for six months or so. The formidable adversary that it was, the cancer once again found a way around.

"Just six months?", you might ask. Was this worth all the trouble? Remember, mesothelioma is aggressive and many patients survive only a year past diagnosis. In contrast, *Siva* was still alive five years past diagnosis. Each move had bought him several months to a year of life. In particular, erlotinib and then the erlotinib–everolimus combination, both suggested by sequencing his cancer genome, had added almost a year and a half of modestly good quality life. A small, albeit solemn, victory for modern medical science.

Imagine, in the future, that doctors have an extended repertoire of drugs to combat cancer. With each countermove the cancer makes, a new move pushes the cancer back to the wall, forcing it to work hard to dodge the bullet. Imagine further, that doctors have the ability to repeatedly check the cancer genome to identify which variants the cancer has evolved as it responds to each move, thus helping guide the next move. Could this game of moves and countermoves be extended much longer? Possibly.

However, cancers, as we have seen, are indeed rather resourceful in finding their way around obstacles. This is particularly true once they achieve a certain momentum in acquiring new variants. Catching them early while they are still slow and sluggish may well be the most important step in this battle.

Of course, that is easier said than done. A tiny clump of cancerous cells rarely announces its presence until it is too late. Imaging techniques like mammograms that can look into our bodies and spot these cancerous clumps usually involve substantial doses of harmful radiation, and also are not very accurate. Much progress needs to be made in this regard. What is promising though is the observation that dying cancer cells shed their genomes into the bloodstream. By drawing a few milliliters of blood from a person and scouring it for genomic fragments carrying key variants, we may be able to spot cancer much earlier than we can today.

Indeed, believes Harold Varmus, ex-director of the United States National Cancer Institute and noted cancer researcher:[166] *I think we can turn most if not all cancers into chronic, controllable, non-lethal disorders, and sometimes cure them, even after they have widely metastasized, if we make the necessary investments in development and testing of new therapies.*

Epilogue

The *Genome* is a fascinating playground where nature, like a child let loose, makes seemingly random changes from one generation to the next—cutting some text here, pasting it there, tweaking the odd character, knocking off some, and inserting others. Much of this casual tinkering we easily take in our stride. On occasion though, things get more challenging, as for *Rom* and *Som*, who passed away at the tender age of one, or for *Dia*, who lost her vision, or for *Sita*, whose blood is severely compromised in its ability to carry oxygen, or for *Ali*, whose infant eyes developed life-threatening tumors.

Frustratingly, even as these ill-effects become apparent, the genome itself remains invisible, concealed deep within the trillions of tiny cells in our bodies, and well beyond the reach of the most powerful microscopes. Modern technology allows us only indirect glimpses, by tearing down our individual genomic books into billions of tiny shreds and elucidating the character sequence in each shred. Then begins the elaborate task of assembling the jigsaw puzzle posed by these shreds and reconstructing the genome. This is an intricate process with lots of nooks and corners and twists and turns, as illustrated in these stories. Nevertheless, it achieves something quite mind-boggling—identification of a few characters from among the billions in the genome that lie at the heart of these ill-effects. It succeeds not always, but often—quite definitively about 50% of the time, and with varying shades of uncertainty another 20% or so of the time, as in the case of *Rom* and *Som*.

Of course, for families at the receiving end of these ill-effects, pinning down the genomic cause is only the first step. Ensuring that future children in the family do not inherit these genomic characters is important, and that is possible today. However, reversing the ill-effects of these characters in living individuals is a much tougher task. That story is in its infancy, but evolving fast. Hopefully, a decade from now, progress on *genome editing* will allow us to *edit* an individual's genome and remove problematic characters early in life, thus keeping their ill-effects at bay.

Then comes the reminder that taming the genome we were born with is not enough, for our genome is dynamic—it changes over our lifetime. The genomes in each of our trillions of cells evolve somewhat independently, yielding trillions of different genomic sequences. Often, one of these trillions of sequences develops a handful of problematic characters, leading to the onset of cancer. What follows is a struggle against the disease, as *Siva*'s ordeal illustrates in the final story.

How does one monitor trillions of genomic sequences to catch any problematic characters early, well before the cancer catches speed? Fortunately, dying cells leak their genomes into the blood, which then finds itself home to the entire cornucopia of genomic variety in our bodies. Hopefully, a decade from now, maybe sooner, a so-called *liquid biopsy* will detect even a slight sprinkling of problematic characters from a blood draw, allowing us to track and treat the cancer well before it builds momentum.

Therefore, and with the hope that the stories in this book will inspire you to devote your energies to the ongoing quest to tame the ill-effects of these quirky genomic characters, we bring this book to an end.

Glossary

Genome. Three billion characters which each of our parents gifts us. With a few exceptions, each cell in our body carries an entire copy of these six billion characters.

Chromosome. These six billion characters are organized into 23 pairs of chromosomes. In each pair, we inherit one member from our father and one from our mother. These two *versions* have slightly different character sequences. The 23rd chromosome pair determines the sex of a person.

Genes. Stretches of characters which serve as recipes for the creation of molecules called proteins. There are roughly 20,000 genes in the genome.

Exons & Introns. Gene recipes are not written in one contiguous stretch in the genome; rather, they are written in various stretches called exons, interrupted by long stretches called introns.

Triplets, Amino Acids & Proteins. Gene recipes in the exons appear as a sequence of character triplets. Each triplet codes for a particular amino acid. The entire recipe thus yields a sequence of amino acids. This sequence of amino acids then acquires a characteristic three-dimensional shape, and is called a protein. This protein participates in various chemical reactions.

Splicing. Before executing a gene recipe, our cells have to identify and remove the introns. This act is called splicing.

Sequencing. The act of reading the genomic character sequence of a given individual is called genome sequencing.

Reference Sequence. This is the aggregate genome sequence of a few healthy individuals. When a person's genome is sequenced, it is compared against the reference sequence to identify differences, and then the focus shifts to these differences.

Read. The act of sequencing the genome of an individual requires chopping up the genome into small pieces. Each piece is called a read.

Read Alignment. The order of these reads in the genome is completely lost in the sequencing process. One indirect way to recover this order is to search for each read in the reference sequence. This search process is called read alignment.

Variant. Differences between a person's genome and the reference sequence are called variants.

Missense. A variant that causes one amino acid to be replaced by another in a gene recipe is called a missense variant.

Nonsense. A variant that causes premature termination of the gene recipe is called a nonsense variant.

Frameshift. A variant that causes jittering of triplets in a gene recipe is called a frameshift variant.

Somatic Variant. Variants we acquire after birth are called somatic variants. These are typically not passed on to our children. They play a key role in converting a normal cell to a cancer cell.

References

1. S. Ishihara. "Tests for colour-blindness". In: *Handaya, Tokyo* (1917).

2. S.S. Deeb. "The molecular basis of variation in human color vision". In: *Clinical Genetics* 67.5 (2005), pp. 369–377.

3. B.C. Verrelli and S.A. Tishkoff. "Signatures of selection and gene conversion associated with human color vision variation". In: *American Journal of Human Genetics* 75.3 (2004), pp. 363–375.

4. S.S. Deeb. "Molecular genetics of color vision deficiencies". In: *Visual Neuroscience* 21.3 (2004), pp. 191–196.

5. International Human Genome Sequencing Consortium. "Finishing the euchromatic sequence of the human genome". In: *Nature* 431 (2004), pp. 931–945.

6. D.M. Hunt et al. "The chemistry of John Dalton's color blindness". In: *Science* 267 (1995), pp. 964–988.

7. J.R. Sparrow, K. Nakanishi, and C.A. Parish. "The lipofuscin fluorophore A2E mediates blue light-induced damage to retinal pigmented epithelial cells". In: *Investigative Ophthalmology and Visual Science* 41.7 (2000), pp. 1981–1989.

8. J.R. Sparrow et al. "A2E, a byproduct of the visual cycle". In: *Vision Research* 43.28 (2003), pp. 2983–2990.

9. S. Schmitz-Valckenberg et al. "Fundus autofluorescence imaging: review and perspectives". In: *Retina* 28.3 (2008), pp. 385–409.

10. S.M. Conley et al. "Increased cone sensitivity to ABCA4 deficiency provides insight into macular vision loss in Stargardt's dystrophy". In: *Biochimica et Biophysica Acta* 1822.7 (2012), pp. 1169–1179.

11. J. Ahn et al. "Functional interaction between the two halves of the photoreceptor-specific ATP binding cassette protein ABCR (*ABCA4*). Evidence for a non-exchangeable ADP in the first nucleotide binding domain". In: *Journal of Biological Chemistry* 278.41 (2003), pp. 39600–39608.

12. N.P. Boyer et al. "Lipofuscin and N-retinylidene-N-retinylethanolamine (A2E) accumulate in retinal pigment epithelium in absence of light exposure: their origin is 11-cis-retinal". In: *Journal of Biological Chemistry* 287.26 (2012), pp. 22276–22286.

13. M. Zhong, L.L. Molday, and S.R. Molday. "Role of the C terminus of the photoreceptor ABCA4 transporter in protein folding, function, and retinal degenerative diseases". In: *Journal of Biological Chemistry* 284 (2009), pp. 3640–3649.

14. R. Battu et al. "Identification of novel mutations in *ABCA4* gene: clinical and genetic analysis of Indian patients with Stargardt's disease". In: *Biomedical Research International* 940864 (2015).

15. M. Lek et al. "Analysis of protein-coding genetic variation in 60,706 humans". In: *Nature* 536.7616 (2016), pp. 285–291.

16. H. Sun, P.M. Smallwood, and J. Nathans. "Biochemical defects in ABCR protein variants associated with human retinopathies". In: *Nature Genetics* 26 (2000), pp. 242–246.

17. T.R. Burke et al. "Retinal phenotypes in patients homozygous for the G1961E mutation in the *ABCA4* gene". In: *Investigative Ophthalmology and Visual Science* 53.8 (2012), pp. 4458–4467.

18. D.M. Lipinski, M. Thake, and R.E. MacLaren. "Clinical applications of retinal gene therapy". In: *Progress in Retinal and Eye Research* 32 (2013), pp. 22–47.

19. J. Kong et al. "Correction of the disease phenotype in the mouse model of Stargardt disease by lentiviral gene therapy". In: *Gene Therapy* 15.19 (2008), pp. 1311–1320.

20. Oxford Biomedica. "Phase I/IIa study of StarGen in patients with Stargardt macular degeneration". In: ***http://clinicaltrials.gov/ct2/show/study/NCT01367444?term=stargen&rank=2*** (2016).

21. S.D. Schwartz et al. "Embryonic stem cell trials for macular degeneration: a preliminary report". In: *The Lancet* 379.9817 (2012), pp. 713–720.

22. Advanced Cell Technology. "Sub-retinal transplantation of hESC derived RPE(MA09-hRPE) cells in patients with Stargardt's macular dystrophy". In: ***http://clinicaltrials.gov/ct2/show/NCT01345006*** (2016).

23. S.W. Peltz et al. "Ataluren as an agent for therapeutic nonsense suppression". In: *Annual Review of Medicine* 64 (2013), pp. 407–425.

24. I.P. Temple et al. "Connexins and the atrioventricular node". In: *Heart Rhythms* 10.2 (2013), pp. 297–304.

25. D. Wanga et al. "Cardiac channelopathy testing in 274 ethnically diverse sudden unexplained deaths". In: *Forensic Science International* 237 (2014), pp. 90–99.

26. B.J. Maron et al. "Prevalence of hypertrophic cardiomyopathy in a general population of young adults: echocardiographic analysis of 4111 subjects in the CARDIA study". In: *Circulation* 92 (1995), pp. 785–789.

27. D.S. Herman et al. "Truncations of titin causing dilated cardiomyopathy". In: *New England Journal of Medicine* 366.7 (2012), pp. 619–628.

28. V. Carmignac et al. "C-terminal titin deletions cause a novel early-onset myopathy with fatal cardiomyopathy". In: *Annals of Neurology* 61.4 (2007), pp. 340–351.

29. R. Knoll et al. "Laminin-alpha4 and integrin-linked kinase mutations cause human cardiomyopathy via simultaneous defects in cardiomyocytes and endothelial cells". In: *Circulation* 116.5 (2007), pp. 515–525.

30. G.I. Gallicano et al. "Desmoplakin is required early in development for assembly of desmosomes and cytoskeletal linkage". In: *Journal of Cellular Biology* 143.7 (1998), pp. 2009–2022.

31. E. Garcia-Gras et al. "Suppression of canonical Wnt/beta-catenin signaling by nuclear plakoglobin recapitulates phenotype of arrhythmogenic right ventricular cardiomyopathy". In: *Journal of Clinical Investigation* 116.7 (2006), pp. 2012–2021.

32. J. Gomes et al. "Electrophysiological abnormalities precede overt structural changes in arrhythmogenic right ventricular cardiomyopathy due to mutations in desmoplakin—A combined murine and human study". In: *European Heart Journal* 33.15 (2012), pp. 1942–1953.

33. G. Quarta et al. "Familial evaluation in arrhythmogenic right ventricular cardiomyopathy: impact of genetics and revised task force criteria". In: *Circulation* 123.23 (2011), pp. 2701–2709.

34. T.J. Pugh et al. "The landscape of genetic variation in dilated cardiomyopathy as surveyed by clinical DNA sequencing". In: *Genetics in Medicine* 16 (2014), pp. 601–608.

35. T.B. Rasmussen et al. "Protein expression studies of desmoplakin mutations in cardiomyopathy patients reveal different molecular disease mechanisms". In: *Clinical Genetics* 84.1 (2013), pp. 20–30.

36. M. Norman et al. "Novel mutation in desmoplakin causes arrhythmogenic left ventricular cardiomyopathy". In: *Circulation* 112 (2005), pp. 636–642.

37. M.J. Landrum et al. "ClinVar: public archive of relationships among sequence variation and human phenotype". In: *Nucleic Acids Research* 42 (2014), pp. 980–985.

38. C.A. James et al. "Exercise increases age-related penetrance and arrhythmic risk in arrhythmogenic right ventricular dysplasia (cardiomyopathy) associated desmosomal mutation carriers". In: *Journal of the American College of Cardiology* 62.14 (2013), pp. 1290–1297.

39. C.L. Lien et al. "Heart repair and regeneration: recent insights from zebrafish studies". In: *Wound Repair and Regeneration* 20.5 (2012), pp. 638–646.

40. B. Loeys et al. "Homozygosity for a missense mutation in fibulin-5 results in a severe form of cutis laxa". In: *Human Molecular Genetics* 11.18 (2002), pp. 2113–2118.

41. D.S. Peabody. "Translation initiation at non-AUG triplets in mammalian cells". In: *Journal of Biological Chemistry* 264.9 (1989), pp. 5031–5035.

42. M. Kozak. "Point mutations define a sequence flanking the AUG initiator codon that modulates translation by eukaryotic ribosomes". In: *Cell* 44.2 (1986), pp. 283–292.

43. H. Liu and L. Wong. "Data mining tools for biological sequences". In: *Journal of Bioinformatics and Computational Biology* 1.1 (2003), pp. 139–167.

44. T. Nakamura et al. "Fibulin-5/DANCE is essential for elastogenesis in vivo". In: *Nature* 415.6868 (2002), pp. 171–175.

45. M. Budatha et al. "Extracellular matrix proteases contribute to progression of pelvic organ prolapse in mice and humans". In: *Journal of Clinical Investigation* 121.5 (2011), pp. 2048–2059.

46. Z. Razinia et al. "Filamins in mechanosensing and signaling". In: *Annual Review of Biophysics* 41 (2012), pp. 227–246.

47. E. Parrini et al. "Periventricular heterotopia: phenotypic heterogeneity and correlation with Filamin A mutations". In: *Brain* 129 (2006), pp. 1892–1906.

48. S. Singh et al. "Case series of 4 infants with severe infantile respiratory failure associated with Filamin A mutation leading to lung transplantation". In: *American Thoracic Society International Conference Abstracts: A106. INTERESTING PEDIATRIC CASES* (2013).

49. S.P. Robertson. "Filamin A: phenotypic diversity". In: *Current Opinion in Genetics and Development* 15.3 (2005), pp. 301–307.

50. A. Verloes et al. “Fronto-otopalatodigital osteodysplasia: clinical evidence for a single entity encompassing Melnick-Needles syndrome, otopalatodigital syndrome types 1 and 2, and frontometaphyseal dysplasia”. In: *Americal Journal of Medical Genetics* 90.5 (2000), pp. 407–422.

51. C. Foley et al. “Expansion of the spectrum of *FLNA* mutations associated with Melnick-Needles syndrome”. In: *Molecular Syndromology* 1.3 (2010), pp. 121–128.

52. K.L. Jones, M.C. Jones, and M. Del Campo. *Smith's recognizable patterns of human malformation.* 2013, p. 762.

53. H.H. Santos et al. “Mutational analysis of two boys with the severe perinatally lethal Melnick-Needles syndrome”. In: *American Journal of Medical Genetics A* 152.3 (2010), pp. 726–731.

54. C. Ris-Stalpers et al. “Substitution of aspartic acid-686 by histidine or asparagine in the human androgen receptor leads to a functionally inactive protein with altered hormone-binding characteristics”. In: *Molecular Endocrinology* 5.10 (1991), pp. 1562–1569.

55. B.A. Kesner et al. “Isoform divergence of the filamin family of proteins”. In: *Molecular Biology and Evolution* 27.2 (2009), pp. 283–295.

56. Y. Feng et al. “*Filamin A (FLNA)* is required for cell–cell contact in vascular development and cardiac morphogenesis”. In: *Proceedings of the National Academy of Sciences* 103.52 (2006), pp. 19836–19841.

57. C.J. Saunders et al. “Rapid whole-genome sequencing for genetic disease diagnosis in neonatal intensive care units”. In: *Science Translational Medicine* 4.154 (2012).

58. R.T. Moon. “Xenopus embryo: beta-catenin and dorsal–ventral axis formation”. In: *eLS. John Wiley and Sons Ltd, Chichester* (2005).

59. S. Yoshiba and H. Hamada. “Roles of cilia, fluid flow, and Ca2+ signaling in breaking of left—right symmetry”. In: *Trends in Genetics* 30.1 (2014), pp. 10–17.

60. W.M. Layton Jr. “Random determination of a developmental process”. In: *The Journal of Heredity* 67 (1976), pp. 336–338.

61. E.C. Oh and N. Katsanis. “Cilia in vertebrate development and disease”. In: *Development* 139.3 (2012), pp. 443–448.

62. C.J. Tabin and K.J. Vogan. “A two-cilia model for vertebrate left–right axis specification”. In: *Genes and Development* 17 (2003), pp. 1–6.

63. S. Sauer and A. Klar. “Left–right symmetry breaking in mice by left–right dynein may occur via a biased chromatid segregation mechanism, without directly involving the Nodal gene”. In: *Frontiers in Oncology* 2.166 (2012).

64. T.P. Yamaguchi. "Heads or tails: Wnts and anterior-posterior patterning". In: *Current Biology* 11.17 (2001), R713–R724.

65. A. Caron, X. Xu, and X. Lin. "Wnt/beta-catenin signaling directly regulates Foxj1 expression and ciliogenesis in zebrafish Kupffer's vesicle". In: *Development* 139.3 (2012), pp. 514–524.

66. J. Deka et al. "Bcl9/Bcl9l are critical for Wnt-mediated regulation of stem cell traits in colon epithelium and adenocarcinomas". In: *Cancer Research* 70.16 (2010), pp. 6619–6628.

67. R. Cole and R. Hariharan. "Approximate string matching: a simpler faster algorithm". In: *SIAM Journal on Computing* 31.6 (2002), pp. 1761–1783.

68. P. Ferragina and G. Manzini. "Indexing compressed text". In: *Journal of the ACM* 52.4 (2005), pp. 552–581.

69. A. Guimier et al. "*MMP21* is mutated in human heterotaxy and is required for normal left–right asymmetry in vertebrates". In: *Nature Genetics* 47.11 (2015), pp. 1260–1263.

70. K. Ahokas et al. "Matrix metalloproteinase-21 is expressed epithelially during development and in cancer and is up-regulated by transforming growth factor-beta1 in keratinocytes". In: *Lab Investigation* 83.12 (2003), pp. 1887–1899.

71. N. Madan et al. "Frequency of beta-thalassemia trait and other hemoglobinopathies in northern and western India". In: *Indian Journal of Human Genetics* 16.1 (2010), pp. 16–25.

72. X. Hong, D.G. Scofield, and M. Lynch. "Intron size, abundance, and distribution within untranslated regions of genes". In: *Molecular Biology and Evolution* 23.12 (2006), pp. 2392–2404.

73. S. Clancy. "RNA splicing: introns, exons and spliceosome". In: *Nature Education* 1.1 (2008), p. 31.

74. K. Gao et al. "Human branch point consensus sequence is yUnAy". In: *Nucleic Acids Research* 36.7 (2008), pp. 2257–2267.

75. A. Corvelo et al. "Genome-wide association between branch point properties and alternative splicing". In: *Public Library of Science, Computational Biology* 6.11 (2010).

76. Z-M. Zheng et al. "Optimization of a weak 3-prime splice site counteracts the function of a bovine papillomavirus type 1 exonic splicing suppressor in vitro and in vivo". In: *Journal of Virology* 74.13 (2000), pp. 5902–5910.

77. M.D. Bashyam et al. "Molecular genetic analyses of beta-thalassemia in South India reveals rare mutations in the beta-globin gene". In: *Journal of Human Genetics* 49 (2004), pp. 408–413.

78. M.D. Bashyam, A.K. Chaudhary, and V. Bhat. "The IVS-II-837 (T>G) appears to be a relatively common 'rare' beta-globin gene mutation in beta-thalassemia patients in Karnataka State, South India". In: *Hemoglobin* 36.5 (2012), pp. 497–503.

79. N.Y. Varawalla, J.M. Old, and D.J. Weatherall. "Rare beta-thalassaemia mutations in Asian Indians". In: *British Journal of Haemotology* 79.4 (1991), pp. 640–644.

80. P. Aldhous. "Bone marrow donors risk DNA identity mix-up". In: *https://www.newscientist.com/article/mg18825234-600-bone-marrow-donors-risk-dna-identity-mix-up* (2005).

81. J. Gaziev and G. Lucarelli. "Hematopoietic stem cell transplantation for thalassemia". In: *Current Stem Cell Research and Therapy* 6.2 (2011), pp. 162–169.

82. G. Lucarelli et al. "Hematopoietic stem cell transplantation in thalassemia and sickle cell anemia". In: *Cold Spring Harbor Perspectives in Medicine* 5.2 (2012).

83. M. Sabloff et al. "HLA-matched sibling bone marrow transplantation for beta-thalassemia major". In: *Blood* 117.5 (2011), pp. 1745–1750.

84. R. Barrangou et al. "CRISPR provides acquired resistance against viruses in prokaryotes". In: *Science* 315.5819 (2007), pp. 1709–1712.

85. M. Jinek et al. "A programmable dual-RNA-guided DNA endonuclease in adaptive bacterial immunity". In: *Science* 337.6096 (2012), pp. 816–821.

86. B. Shen et al. "Efficient genome modification by CRISPR-Cas9 nickase with minimal off-target effects". In: *Nature Methods* 11 (2014), pp. 399–402.

87. L. Cong et al. "Multiplex genome engineering using CRISPR/Cas systems". In: *Science* 339.6121 (2013), pp. 819–823.

88. H. Yin et al. "Genome editing with Cas9 in adult mice corrects a disease mutation and phenotype". In: *Nature Biotechnology* 32 (2014), pp. 551–553.

89. J. Lewis et al. "A common human beta globin splicing mutation modeled in mice". In: *Blood* 91.6 (1998), pp. 2152–2156.

90. J. Morley Smith. "Red reflex image, in public domain". In: *http://upload.wikimedia.org/wikipedia/commons/d/d8/Rb_whiteeye.PNG* (2008).

91. A. Mallipatna. "White reflex image, Copyright: Narayana Nethralaya". In: *Private Communication* (2015).

92. H. Vyas. "Photographer's awareness saved girl's vision". In: *http://timesofindia.indiatimes.com/city/bengaluru/Photographers-awareness-saved-girls-vision/articleshow/13160469.cms* (2012).

93. A. Knudson. "Mutation and cancer: statistical study of retinoblastoma". In: *Proceedings of the National Academy of Sciences* 68.4 (1971), pp. 820–823.

94. A.M. Narasimha et al. "Cyclin D activates the Rb tumor suppressor by mono-phosphorylation". In: *ELife* 3 (2014).

95. C.R. de Andrade et al. "A molecular study of first and second RB1 mutational hits in retinoblastoma patients". In: *Cancer Genetics and Cytogenetics* 167 (2006), pp. 43–46.

96. W. Chen and S. Jinks-Robertson. "The role of the mismatch repair machinery in regulating mitotic and meiotic recombination between diverged sequences in yeast". In: *Genetics* 151 (1999), pp. 1299–1313.

97. M.C. LaFave and J. Sekelsky. "Mitotic recombination: why? when? how? where?" In: *Public Library of Science, Genetics* 5.3 (2009).

98. W. Jiao et al. "Aberrant nucleocytoplasmic localization of the retinoblastoma tumor suppressor protein in human cancer correlates with moderate/poor tumor differentiation". In: *Oncogene* 27 (2008), pp. 3156–3164.

99. E. Zacksenhaus et al. "A bipartite nuclear localization signal in the retinoblastoma gene product and its importance for biological activity". In: *Molecular and Cellular Biology* 13.4 (1993), pp. 4588–4599.

100. A. MacCarthy et al. "Non-ocular tumours following retinoblastoma in Great Britain 1951 to 2004". In: *British Journal of Ophthalmology* 93.9 (2009), pp. 1159–1162.

101. A. Turing. "The chemical basis of morphogenesis". In: *Philosophical Transactions of the Royal Society of London, Series B* 237.641 (1952).

102. B.L. Gallie et al. "Developmental basis of retinal-specific induction of cancer by *RB* mutation". In: *Cancer Research* 59.7 Suppl (1999), 1731s–1735s.

103. X.L. Xu et al. "Rb suppresses human cone-precursor-derived retinoblastoma tumours". In: *Nature* 514.7522 (2014), pp. 385–388.

104. A. Balmer, L. Zografos, and F. Munier. "Diagnosis and current management of retinoblastoma". In: *Oncogene* 25.38 (2006), pp. 5341–5349.

105. B.P. Alter et al. "Cancer in Fanconi anemia". In: *Blood* 101.5 (2003), pp. 2072–2073.

106. D.W. Smithers. "Family histories of 459 patients with cancer of the breast". In: *British Journal of Cancer* 2.2 (1948), pp. 163–167.

107. E.B. Claus, N.J. Risch, and W.D. Thompson. "Age at onset as an indicator of familial risk of breast cancer". In: *American Journal of Epidemiology* 131.6 (1990), pp. 961–972.

108. J.M. Hall et al. "Linkage of early-onset familial breast cancer to chromosome 17q21". In: *Science* 250.4988 (1990), pp. 1684–1689.

109. R. Wooster et al. "Identification of the breast cancer susceptibility gene *BRCA2*". In: *Nature* 378 (1995), pp. 789–792.

110. A. Jolie. "My medical choice". In: *http://www.nytimes.com/2013/05/14/opinion/my-medical-choice.html* (2013).

111. A. Jolie Pitt. "Angelina Jolie Pitt: Diary of a surgery". In: *http://www.nytimes.com/2015/03/24/opinion/angelina-jolie-pitt-diary-of-a-surgery.html* (2015).

112. D.A. Kreutzer and J.M. Essigmann. "Oxidized, deaminated cytosines are a source of C->T transitions in vivo". In: *Proceedings of the National Academy of Sciences* 95.7 (1998), pp. 3578–3582.

113. A. Valavanidis, T. Vlachogianni, and K. Fiotakis. "Tobacco smoke: involvement of reactive oxygen species and stable free radicals in mechanisms of oxidative damage, carcinogenesis and synergistic effects with other respirable particles". In: *International Journal of Environmental Research and Public Health* 6.2 (2009), pp. 445–462.

114. D.E. Heck et al. "UVB light stimulates production of reactive oxygen species". In: *The Journal of Biological Chemistry* 278.25 (2003), pp. 22432–22436.

115. Y. Huang and L. Li. "DNA crosslinking damage and cancer—a tale of friend and foe". In: *Translational Cancer Research* 2.3 (2013), pp. 144–154.

116. G.L. Moldovan and A.D. D'Andrea. "How the Fanconi anemia pathway guards the genome". In: *Annual Review of Genetics* 43 (2009), pp. 223–249.

117. M. Huang et al. "Human MutS and FANCM complexes function as redundant DNA damage sensors in the Fanconi Anemia pathway". In: *DNA Repair* 10.12 (2011), pp. 1203–1212.

118. A.D. D'Andrea. "BRCA1: A missing link in the Fanconi anemia/BRCA pathway". In: *Cancer Discovery* 3.4 (2013), pp. 376–378.

119. G. Ast. "How did alternative splicing evolve". In: *Nature Reviews Genetics* 5.10 (2004), pp. 773–782.

120. E. Buratti et al. “Aberrant 5’ splice sites in human disease genes: Mutation pattern, nucleotide structure and comparison of computational tools that predict their utilization”. In: *Nucleic Acids Research* 35.13 (2007), pp. 4250–4263.

121. X. Roca, R. Sachidanandam, and A.R. Krainer. “Intrinsic differences between authentic and cryptic 5’ splice sites”. In: *Nucleic Acids Research* 31.21 (2003), pp. 6321–6333.

122. M.J. Hicks et al. “Competing upstream 5’ splice sites enhance the rate of proximal splicing”. In: *Molecular and Cellular Biology* 30.8 (2010), pp. 1878–1886.

123. C. Attanasio, A. David, and M. Neerman-Arbez. “Outcome of donor splice site mutations accounting for congenital afibrinogenemia reflects order of intron removal in the fibrinogen alpha gene (*FGA*)”. In: *Blood* 101.5 (2003), pp. 1851–1856.

124. S.C. Chandrasekharappa et al. “Massively parallel sequencing, aCGH, and RNA-Seq technologies provide a comprehensive molecular diagnosis of Fanconi anemia”. In: *Blood* 121.22 (2013), pp. 138–148.

125. Y.J. Machida et al. “UBE2T is the E2 in the Fanconi anemia pathway and undergoes negative autoregulation”. In: *Molecular Cell* 23.4 (2006), pp. 589–596.

126. A.F. Alpi et al. “Mechanistic insight into site-restricted monoubiquitination of FANCD2 by Ube2t, FANCL, and FANCI”. In: *Molecular Cell* 32.6 (2008), pp. 767–777.

127. A.J. Deans and S.C. West. “DNA interstrand crosslink repair and cancer”. In: *Nature Reviews* 11.7 (2011), pp. 467–480.

128. R. Ceccaldi et al. “Bone marrow failure in Fanconi anemia is triggered by an exacerbated p53-p21 DNA damage response that impairs hematopoietic stem and progenitor cells”. In: *Cell Stem Cell* 11.1 (2012), pp. 36–49.

129. R. Roy, J. Chun, and S.N. Powell. “*BRCA1* and *BRCA2*: different roles in a common pathway of genome protection”. In: *Nature Reviews* 12.1 (2011), pp. 68–78.

130. R.L. Attanoos. “Malignant mesothelioma: Asbestos exposure”. In: *Occupational Cancers* (2014), pp. 273–284.

131. Cancer Research UK. “Mesothelioma risks and causes”. In: *http://www.cancerresearchuk.org/about-cancer/type/mesothelioma/about/mesothelioma-risks-and-causes* (2016).

132. M. Krishnan and S.G. Ray. "Banned in 52 countries, asbestos is India's next big killer". In: *http://archive.tehelka.com/story_main46.asp?filename=Cr070810banned.asp* (2010).

133. R.A. Weiss and P.K. Vogt. "100 years of Rous sarcoma virus". In: *Journal of Experimental Medicine* 208.12 (2011), pp. 2351–2355.

134. P. Rous. "A sarcoma of the fowl transmissible by an agent separable from the tumor cells". In: *Journal of Experimental Medicine* 13.4 (1911), pp. 397–411.

135. A.P. Czernilofsky et al. "Nucleotide sequence of an avian sarcoma virus oncogene (src) and proposed amino acid sequence for gene product". In: *Nature* 287.5779 (1980), pp. 198–203.

136. D.M. Parkin. "The global health burden of infection-associated cancers in the year 2002". In: *International Journal of Cancer* 118.12 (2006), pp. 3030–3044.

137. J.S. Butel and J.A. Lednicky. "Cell and molecular biology of simian virus 40: Implications for human infections and disease". In: *Journal of the National Cancer Institute* 91.2 (1999), pp. 119–134.

138. A.L. Cleaver et al. "Long-term exposure of mesothelial cells to SV40 and asbestos leads to malignant transformation and chemotherapy resistance". In: *Carcinogenesis* 35.2 (2014), pp. 407–414.

139. M. Carbone. "Simian virus 40 and human tumors: it is time to study mechanisms". In: *Journal of Cellular Biochemistry* 76.2 (1999), pp. 189–193.

140. L. Zhang et al. "Tissue tropism of SV40 transformation of human cells: role of the viral regulatory region and of cellular oncogenes". In: *Genes and Cancer* 1.10 (2010), pp. 1008–1020.

141. C. Tomasetti and B. Vogelstein. "Variation in cancer risk among tissues can be explained by the number of stem cell divisions". In: *Science* 347.6217 (2015), pp. 78–81.

142. S. Papaspyros and S. Papaspyros. "Surgical management of malignant pleural mesothelioma: impact of surgery on survival and quality of life—relation to chemotherapy, radiotherapy, and alternative therapies". In: *International Scholarly Research Notes, Surgery* 2014 (2014).

143. D. Hanahan and R. A. Weinberg. "Hallmarks of cancer: the next generation". In: *Cell* 144.5 (2011), pp. 646–674.

144. B. Vogelstein et al. "Cancer genome landscapes". In: *Science* 339.6127 (2013), pp. 1546–1558.

145. National Comprehensive Cancer Network. "Malignant pleural mesothelioma". In: `http://www.nccn.org/patients/guidelines/mpm/` (2014).

146. N.J. Vogelzang et al. "Phase III study of pemetrexed in combination with cisplatin versus cisplatin alone in patients with malignant pleural mesothelioma". In: *Journal of Clinical Oncology* 21.14 (2003), pp. 2636–2644.

147. D-W. Shen et al. "Cisplatin resistance: a cellular self-defense mechanism resulting from multiple epigenetic and genetic changes". In: *Pharmacological Reviews* 64.3 (2012), pp. 706–721.

148. B. Castagneto et al. "Phase II study of pemetrexed in combination with carboplatin in patients with malignant pleural mesothelioma (MPM)". In: *Annals of Oncology* 19.2 (2008), pp. 370–373.

149. W. DeNeve et al. "Discrepancy between cytotoxicity and DNA interstrand crosslinking of carboplatin and cisplatin in vivo". In: *Investigational New Drugs* 8.1 (1990), pp. 17–24.

150. A. K. Nowak et al. "A multicentre phase II study of cisplatin and gemcitabine for malignant mesothelioma". In: *British Journal of Cancer* 87.5 (2002), pp. 491–496.

151. E. Mini et al. "Cellular pharmacology of gemcitabine". In: *Annals of Oncology* 17.5 (2006), pp. 7–12.

152. A. Efeyan and M. Serrano. "p53: guardian of the genome and policeman of the oncogenes". In: *Cell Cycle* 6.9 (2007), pp. 1006–1010.

153. L.M. Abegglen et al. "Potential mechanisms for cancer resistance in elephants and comparative cellular response to DNA damage in humans". In: *Journal of the Americal Medical Association* 314.17 (2015), pp. 1850–1860.

154. K. Schlereth et al. "Characterization of the p53 cistrome–DNA binding cooperativity dissects p53's tumor suppressor functions". In: *Public Library of Science, Genetics* 9.8 (2013).

155. L.R. Dearth et al. "Inactive full-length p53 mutants lacking dominant wild-type p53 inhibition highlight loss of heterozygosity as an important aspect of p53 status in human cancers". In: *Carcinogenesis* 28.2 (2007), pp. 289–298.

156. A.C. Joerger and A.R. Fersht. "The tumor suppressor p53: from structures to drug discovery". In: *Cold Spring Harbor Perspectives in Biology* 2.6 (2010).

157. U. ManChon et al. "Prediction and prioritization of rare oncogenic mutations in the cancer kinome using novel features and multiple classifiers". In: *Public Library of Science, Computational Biology* 10.4 (2014).

158. F.A. Shepherd et al. "Erlotinib in previously treated non-small-cell lung cancer". In: *New England Journal of Medicine* 353.2 (2005), pp. 123–132.

159. C-Q. Zhu et al. "Role of *KRAS* and *EGFR* as biomarkers of response to Erlotinib in National Cancer Institute of Canada clinical trials group study BR.21". In: *Journal of Clinical Oncology* 26.26 (2008), pp. 4268–4275.

160. L.L. Garland et al. "Phase II study of erlotinib in patients with malignant pleural mesothelioma: a southwest oncology group study". In: *Journal of Clinical Oncology* 25.17 (2007), pp. 2406–2413.

161. "Erlotinib in treating patients with malignant mesothelioma of the lung". In: *https://clinicaltrials.gov/ct2/show/NCT00039182* (2013).

162. S. Mori et al. "The mTOR pathway controls cell proliferation by regulating the FoxO3a transcription factor via SGK1 Kinase". In: *Public Library of Science One* 9.2 (2014).

163. S-Y. Han et al. "Functional evaluation of *PTEN* missense mutations using in-vitro phosphoinositide phosphatase assay". In: *Cancer Research* 60.12 (2000), pp. 3147–3151.

164. G. Iyer et al. "Genome sequencing identifies a basis for everolimus sensitivity". In: *Science* 338.6104 (2012), p. 221.

165. V.A. Papadimitrakopoulou et al. "Everolimus and erlotinib as second- or third-line therapy in patients with advanced non-small-cell lung cancer". In: *Journal of Thoracic Oncology* 7.10 (2012), pp. 1594–1601.

166. H. Varmus. "Can cancer be stopped?" In: *http://www.nydailynews.com/opinion/harold-varmus-cancer-stopped-article-1.2181020* (2015).

167. M. Legrand. "The Legrand orange book". In: *http://www.latextemplates.com/template/the-legrand-orange-book* (2016).

168. K. Goyal. "The Calibre project". In: *https://calibre-ebook.com/about#history* (2016).

169. FreeDesignFile. "Flower background free vector". In: *http://all-free-download.com/free-vector/download/flower-background_532336.html* (2016).

Index

Acknowledgments

Three worlds meet in this book—the worlds of clinical practice, molecular biology, and computer algorithms. While fascinating, the interplay between these worlds necessarily pushes us into zones well outside our individual expertise. The stories in this book therefore reflect the combined efforts of several individuals from all three worlds, anchored by our team at *Strand Life Sciences*.

Dr. Meenakshi Bhat, Clinical Geneticist and Professor at the Center for Human Genetics, Bangalore, provided us with the first, and the most inspiring, introduction to the clinical world. The team at Strand cut its teeth working with Dr. Bhat. Indeed, Story 4 describes one of the cases from our collaboration with Dr. Bhat.

Dr. Rajani Battu, Ophthalmologist and Retina Specialist at Narayana Nethralaya, Bangalore, worked with us extensively. The workshops she organized gave us the most fascinating introduction to the world of retinal dystrophies. Story 2 describes one of our first cases in collaboration with Dr. Battu.

Dr. Subhash Chandra, Cardiologist at Vikram Hospital, Bangalore, and a Fellow of the Royal College of Physicians, gave us our introduction to cardiology. Story 3 describes one of our cases in collaboration with Dr. Chandra.

Prof. Stephen Kingsmore, Professor of Pediatrics, University of Missouri-Kansas City School of Medicine, and a Fellow of the Royal College of Pathologists, was a serendipitous connection. Story 5 is based on his patient(s). This is the only story here for which genome sequencing data was generated outside the laboratories at Strand. A collaboration with *Illumina*, the company whose genome sequencing hardware enabled all our stories, gave us access to this data. This story recounts our analysis of this data and the discovery of a new gene. Prof. Kingsmore's group discovered this gene independently from the same data and published this discovery along with several collaborators. Prof. Kingsmore is one of the pioneers of genomic diagnosis in pediatric medicine, so it was an honor for us to compare notes on this case.

Dr. Sunil Bhat, Consultant in Pediatric Hematology, Oncology and Bone Marrow Transplantation at Narayana Hrudayalaya, Bangalore, referred the patient described in Story 6.

Dr. Ashwin Mallipatna, Pediatric Ophthalmologist and Retinoblastoma Specialist at Narayana Nethralaya, Bangalore, gave us a fascinating introduction to Retinoblastoma. He also worked with us on optimizing our measurement techniques. Story 7 describes one of the cases we collaborated on.

Dr. Annie Hasan, Senior Consultant in Genetics, Kamineni Hospitals, Hyderabad, has worked with us on several cases. A few of these have been eye-openers; an extra trick or two was needed to obtain the right diagnoses in these cases. Story 8 describes one such case.

Dr. Amit Verma, Consultant in Molecular Oncology and Cancer Genetics, Max Healthcare, New Delhi, has worked with us on several cancer cases. Story 9 describes one such case.

My gratitude to all the above clinicians for working with us and for their consent to include these stories in this book.

Erica Ramos, *Elliot Margulies*, *Sean Humphray* and *John Peden*, all at *Illumina*, were instrumental in giving us access to the data in Story 5 and helped in drawing some of the conclusions.

A huge team at *Strand Life Sciences* provided the foundation for these stories. A few names are listed below. My gratitude to all others at Strand who have not been explicitly listed; this book stands on your shoulders. The stories presented here are but a few of the numerous cases that we've worked on together, touching many lives in the process.

Vaijayanti Gupta, *Binay Panda*, *Preveen Ramamoorthy* and *Satish Sankaran* led the molecular biology laboratories in which genomes of most of the patients in these stories were deciphered.

Vamsi Veeramachaneni, *Shanmukh Katragadda*, *Bhupender Singh*, *Rohit Gupta*, *Radhakrishna Bettadapura* and the teams they led designed computer algorithms to distill the vast amounts of data generated by our molecular biology laboratories.

Arunabha Ghosh, *Anand Janakiraman*, *Nimisha Gupta*, *Sujaya Srinivasan* and the teams they led built the software interfaces that brought all relevant scientific literature to our fingertips, thus enabling our hunt for the genomic characters in these stories. *R. Prabhakar* and his team ensured that the supporting IT systems ran reliably and responsively.

Shuba Krishna, *Smita Agarwal*, *Ashraf Mannan*, *Urvashi Bahadur*, *Rupali Gadkari*, *Jemima Jacob*, *Aparna Ganapathy* and the teams they led were the detectives who worked on these and several other cases, often scouring through reams of information to fish out one offending character among six billion.

Jyoti Bajpai, *Prasanna Shirol*, *Payal Manek*, and *Khaleel Ahmed* interfaced with several of the clinicians involved in these stories. *Prasanna Shirol*'s personal battle with one such genomic character and his untiring activism in spreading awareness and helping others is one of the inspirations behind this book.

Achintya Das and *Nilesh Tiwari* helped with modeling structural changes on account of some of the offending characters described in these stories. *Suman Kapoor* helped reconfirm some of the genomic phenomena discovered in these stories using alternative experiments.

Finally, *Kas Subramanian*, *Thiru Reddy*, and *Vijay Chandru* provided overall cover for all of this, financially, as well as through their relationship with clinicians and hospitals. Thanks specifically to *Vijay Chandru* for all the encouragement he's provided over the years.

Several people read and commented on early versions of these stories. *Sowmya Raghavan*, my wife and seasoned molecular biologist, was invariably the first to read and comment on each story. *Ananya Ramgopal*, an eleventh-grade student, was my benchmark reader. Her comments helped calibrate the presentation for readability. *Devanshi Gupta* and *Roshni Sudan*, both high school students, provided similar feedback. *Gautham Machiraju*, an undergraduate student, read the text very carefully and helped edit it for better readability. *Maria Kammerer*'s sharp eye was extremely helpful in catching minute errors in the text.

Detailed comments by *Rajani Battu*, *Babu Narayanan*, and *Manoj Varma*, all seasoned practitioners in their respective fields, provided great encouragement. *Pankaj Agarwal*, an experienced computational biologist, read the manuscript and, as on numerous occasions in the past, offered all support and encouragement, for which I am very grateful. *Prof. Sachin Maheshwari*, under whose tutelage several generations of computer scientists have graduated, offered the most in-depth and sharpest criticism, including the sage advice to minimize algorithmic and procedural descriptions. This has, no doubt, made the text far more free-flowing than it otherwise would be. *Prof. D. Balasubramanian*, an eminent scientist whose science columns in *The Hindu* are quite popular, gave very encouraging feedback as well. *Prof. Dipankar Chatterji*, another eminent scientist and until recently president of the Indian Academy of Sciences, also offered very useful critical feedback.

I am particularly grateful to *Narasimhan Nagan*, Fellow of the American College of Medical Genetics and Genomics (ACMG), for his detailed comments and appreciation following a thorough and expeditious review of a draft version of this book, performed in his personal capacity.

In addition, *Kannan Padmanabhan*, *Maya Malpani*, *Sandeep Tyagi*, *Rajesh Sundaresan*, *Swaroop Aradhya*, *Namitha Kumar*, *Rebecca Manohar*, *Naveen Sivadasan*, *Rema Sivadas*, *Namitha Kumar*, *Padmaja Veeramachaneni*, *Jhelam Deshpande*, *Nitin Deshpande*, and *Swapnesh Banerjee* provided helpful comments. Several students at the *Asian College of Journalism* in Chennai read a chapter or two as part of a brief course on science journalism and wrote reviews, some of which were quite interesting. Comments from all the above have helped to improve the presentation of these stories significantly.

By the time I reached the home stretch of this book, my senses were saturated on account of repeated reading and rewriting. A final pass was needed to ensure no embarrassing errors remained. Help came in the form of several volunteers at Strand who helped proofread the final version: *Sudhir Borgonha*, *Maria Kammerer*, *Bhargavi Ramesh*, *Aditi Aggarwal*, *Shreya Paliwal*, *Suhasini Singh*, *Sunil Cherukuri*, *R. Durairaj*, *Smita Agrawal*, *Urvashi Bahadur* and *Nishita Thota*. *Anand Janakiraman* gave the draft a particularly thorough reading and provided several incisive comments.

Creative ideas for the cover came from *Nishi Gandha*, *Nishita Thota*, *Suman Kapoor*, *Irene Patric* and *Anand Janakiraman*. The idea to use flowers on the cover came from *Anand Janakiraman*. Flowers come in a variety of colors driven by genomic variation. Moreover, they remind us that genomic variation is beautiful and delicate in all its forms, hence the cover.

Typesetting these stories in a presentable way required help from several open-source initiatives based on the LaTeXplatform. It takes a lot of technical work to produce an aesthetic print book and e-book from the same source files, and I cannot imagine another platform that produces such quality while providing exquisite control to the author, all for free. The *Legrand Orange* LaTeX book style by *Mathias Legrand*[167] served as an excellent starting point for the internals of this book. The *Calibre* project founded by *Kovid Goyal*[168] was invaluable in the last mile of creating the e-book. *FreeDesignFile*[169] provided some interesting flower graphics that helped set a base for designing the cover.

The *Indian Institute of Science* press came forward offering to co-publish this book along with Strand Life Sciences. Both press committee chairs, *Prof. G.K. Ananthasuresh* and *Prof. Amaresh Chakrabarti*, were very supportive of a flexible arrangement that would allow both sides to work to their respective strengths. Copy editing by *Sreenivasa Rao* served to provide the final polish to the manuscript. *Kavitha Harish* was very helpful in overseeing the entire process.

My gratitude to all these people. In particular, special thanks to the patients and their families for allowing their stories to be represented here.

Much of this writing was done at home, and did take away from valuable family time. My gratitude to my family for bearing simultaneously with my physical presence and my mental absence. Hopefully, that is about to change.

Finally, my thanks to you, the reader, for taking this book in hand, even if for a quick browse and not an extensive read. Much needs to be done to quell the impact of these genomic characters on patients in general. Hopefully, these stories will inspire you to devote your energies to this cause.

About the Author

Ramesh Hariharan

Ramesh Hariharan is a computer scientist, bioinformatician, and entrepreneur. He is currently the Chief Technology Officer at Strand Life Sciences, and an Adjunct Professor at the Indian Institute of Science (IISc). Strand Life Sciences was founded in the year 2000 as a spin-off from IISc by Ramesh and three colleagues who were fascinated by the use of computer algorithms in biology and medicine. The stories in this book come from first-hand experience at Strand. Ramesh is a fellow of the Indian Academy of Sciences and the Indian National Academy of Engineering, and a recipient of the first Dewang Mehta Award for Innovation in Information Technology awarded by the Government of India. He was also selected to the MIT Technology Review's top-100 innovators list in 2002. Ramesh lives in Bengaluru with his wife, two children, and parents.

www.ingramcontent.com/pod-product-compliance
Ingram Content Group UK Ltd.
Pitfield, Milton Keynes, MK11 3LW, UK
UKHW060736290826
14794UKWH00019B/232/J

9 789352 657438